AIRLIFE'S AIRLINERS 13

BOEING 757

AIRLIFE'S AIRLINERS 13

BOEING 757

Philip Birtles

Airlife
England

Copyright © 2000 Philip Birtles

First published in the UK in 2000
by Airlife Publishing Ltd

British Library Cataloguing-in-Publication Data
A catalogue record for this book
is available from the British Library

ISBN 1 84037 918 5

The information in this book is true and complete to the best of our knowledge. All recommendations are made without any guarantee on the part of the Publisher, who also disclaims any liability incurred in connection with the use of this data or specific details.

All rights reserved. No part of this book may be reproduced or transmitted in any form or by any means, electronic or mechanical including photocopying, recording or by any information storage and retrieval system, without permission from the Publisher in writing.

Printed in Singapore by Kyodo Printing Co (S'pore) Pte Ltd

Airlife Publishing Ltd
101 Longden Road, Shrewsbury, SY3 9EB, England
E-mail: airlife@airlifebooks.com
Website: www.airlifebooks.com

BELOW: The Boeing 757 has been very popular with the world's airlines with orders for 922 aircraft and 810 delivered by the end of July 1998. Northwest 757-200 N536US is seen at Detroit in May 1998. *Philip Birtles*

CONTENTS

INTRODUCTION	6
1. EVOLUTION	8
2. DESIGN AND DEVELOPMENT	16
3. PRODUCTION	32
4. TECHNICAL SPECIFICATION	38
5. IN SERVICE	48
6. CUSTOMERS	66
7. ACCIDENTS AND INCIDENTS	102
8. PRODUCTION LIST	104
9. CHRONOLOGY	124
INDEX	128

INTRODUCTION

Production of the Boeing 757 is now approaching 1,000 aircraft with very little modification until the recent fuselage stretch with the 757-300. The aircraft has become an industry standard, particularly with continental scheduled operations and with a number of major charter operators in North America and Europe. Although some 757s are operating in Asia, due to the substantial number of people wishing to fly in that region, larger aircraft are more popular. In addition to the passenger versions, Boeing has managed to produce a cost effective small cargo/package transport, which has been ordered in significant numbers by UPS as the 757 Package Freighter. The size of aircraft allows high frequency overnight services to be provided between national hubs, and the final destinations.

The 757 has a first generation two-pilot digital EFIS flight deck with cathode-ray tube displays. Power is provided by Rolls-Royce RB.211-535 or Pratt & Whitney PW2000 turbofans, which give quiet, economic and cost effective operation. The aircraft is pleasant to handle both on the ground and in the air, and maintenance is straightforward.

Boeing introduced the stretched 757-300 to attempt to revive lagging sales, but to date the new version has not attracted significant orders. Another possible development is a longer range version of the standard 757-200 to cater for the longer thin routes where a wide body airliner would be uneconomic.

British Airways have been operating 757s on intensive European routes since the first deliveries in January 1983, but has now come to an agreement with Boeing and DHL to dispose of the majority of its fleet for conversion to freighters. The first 757 is expected to leave BA service in August 2000, with a number of services being taken over by smaller Boeing 737s and, increasingly, by the Airbus A320 family.

INTRODUCTION

In addition to the normal desk-top research, there has been significant help from a number of sources, including many of the operators. Special thanks are due to Russell Ison, Media Relations Manager; my old colleague Brian Portch, General Manager Finance & Systems; Malcolm Crawford, Project Manager Group Operations; Captain Mike McKern, Assistant to Fleet Manager operations, Peter Fraser, Group Leader Crew Training Unit – all at Britannia Airways – and to the many others at the airline who have been particularly helpful in providing an overview of 757 operations.

With any book the first task is to contact all the operators to request photos and details of the fleet and route structure. Liz Fang and her colleagues at HSBC in Hong Kong translated my letters to the Chinese operators, but even that did not achieve any replies. Where there were gaps caused by lack of response from the airlines, I was lucky to make contact with David Riley of Asian Aviation Photography and Gary Tahir in Canada who have provided some excellent pictures. My regular friend, Nick Granger, managed to fill a number of gaps with his excellent photos and my old friend Peter Crossley helped with material from Boeing. Without the enthusiastic help from these people, the book would not achieve the high quality expected.

The basis of the production table is taken from the excellent *Jet Airliner Production List Volume 1 – Boeing*, compiled by John Roach and A E Eastwood with additions from a number of sources including *Air Letter*.

As far as possible all the material used in this book is correct at the time of going to press and any errors are entirely the responsibility of the compiler.

Philip Birtles
Stevenage, Herts

1 Evolution

With overall sales now well past the 900 mark, comprised mostly of the basic Series 200 model, the 757 has been a successful programme, devoid of the cost of additional development. At a time when the evolution of a modern airliner can take up to 10 years, demanding substantial financial and engineering resources — and often requiring the services of a crystal ball — the 757 was certainly the right aircraft at the right time. When looking into the expected life of an airliner, and any subsequent developments, new technology can be introduced if it is cost-effective — but only if it is suitable for inclusion without major changes. Fortunately, as the design of modern aircraft is increasingly controlled by computers, improvements can often be made by changes in the software to further enhance the operation of the aircraft.

It takes a company the size of Boeing, with an annual turnover greater than the gross national product of some reasonable sized countries, to be able to develop a full range of airliners. If the development of a new type is not underwritten by the series production of other models, even Boeing can find its finances stretched. In the 1970s, with the onset of the oil

crisis that pushed up fuel prices dramatically, airlines saw their operating costs spiral almost out of control overnight. This reduction in the fortunes of the airlines had a major knock-on effect on the aerospace manufacturers. Boeing was forced to cut its work force by two thirds, to 36,000, over a period of three years to ensure its own survival. However, with the end of the 1970s and the return to improved fortunes, the workforce was able to return to some 75,000.

It was in the depths of this economic crisis that the Boeing Commercial Airplane Group at Seattle set up the New Aircraft Program. The aim was to predict the needs of the airline market over the next 10 to 15 years to replace (or augment) the Model 707, 727, 737 and 747 which were still in production but, in the case of the first two, were approaching the end of their construction cycle. If appropriate, it was necessary to consider a family of new airliners maintaining as much commonality as possible to reduce development, production and maintenance costs.

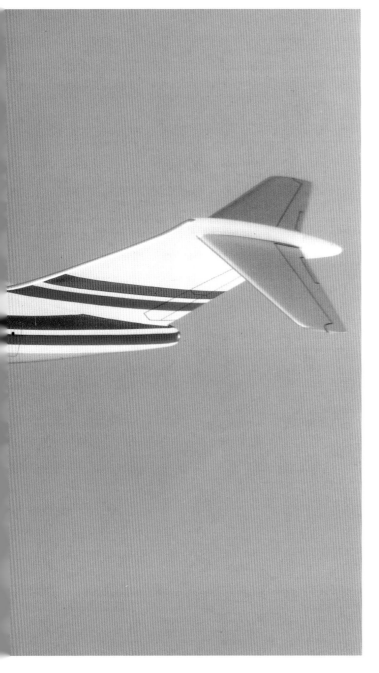

By the mid-1970s, to keep costs down, Boeing was investigating a stretch of the 727 and 737 — and how to overcome the inflationary spiral associated with the development of a totally new airliner. The projected 727-300 'stretch' was first revealed in early 1974 and shown in model form at the Paris Show in June 1975. It was offered in two forms, the minimum change 727-300A and the stretched 727-300B with room for 189 passengers. The three 89kN (20,000lb) thrust Pratt & Whitney (P&W) JT8D-217 engines would take the aircraft over stage lengths of 3,700km (2,000nm). United Airlines had shown an interest in a more advanced 727-300B, but was reluctant to place an order for an aircraft which did not feature the quieter and more fuel-efficient turbofans.

THE 7N7 EMERGES

A year later the project had adopted a 40–50 percent commonality with the 737 airframe and the designation 7N7, under the Boeing design number 761. The major change was a new advanced technology wing attached to a stretched fuselage with power coming from a pair of wing mounted podded CFM-56 or JT10 engines giving a range of over 3,700km (2,000nm). Two fuselage lengths were on offer, either to accommodate 154 or 169 passengers, and there was the option of two or three flight crew operation.

By the middle of 1977, the 7N7 had evolved into a largely new airframe with a T-tail and two wing mounted engines bringing it to 20 seats less than the projected larger wide bodied 7X7. The T-tail layout was chosen virtually by accident. When Boeing tunnel-tested the basic 727 empennage without the centre engine duct, the reduced drag was found to save two per cent of fuel consumption. Furthermore, this added no extra weight and did not cause superstall problems. With United having rejected the 727-300B project, Boeing was evolving the 7N7 to meet this still to be satisfied requirement. The passenger capacity ranged between 160 and 180 seats and powerplants being proposed included the P&W JT10D4s, or cropped fan Rolls-Royce (R-R) RB.211-535s and General Electric (GE) CF6-32s.

Despite the increased capacity, 7N7 operating economy depended on low structural weight with little increase over the 727-200. To keep down both weight and costs, the range was set at relatively short 1,300km (700nm) sectors giving one-stop domestic US coast-to-coast operations, but the generous wing area allowed room for extra fuel for longer European routes, and for improved hot and high performance. Meanwhile, studies were continuing of developments of both the 727 and 737.

Within two or three months, the 7N7 had changed radically, abandoning the traditional six abreast cabin width of the earlier Boeing jet airliners, now increased to a similar size to the projected 7X7 — with a seven abreast seating twin-aisle layout. The new design was becoming a leading contender for the US

LEFT: As the Boeing 757 moved through many 7N7 initial design stages, its size and internal configuration changed several times. Throughout this process, the aircraft retained the T-tailed configuration inherited from the Model 727, but began to adopt more features of the Model 737 — such as the twin, underwing engines. This model shows what Boeing's 7N7 design team were thinking around 1977.
The Aviation Picture Library

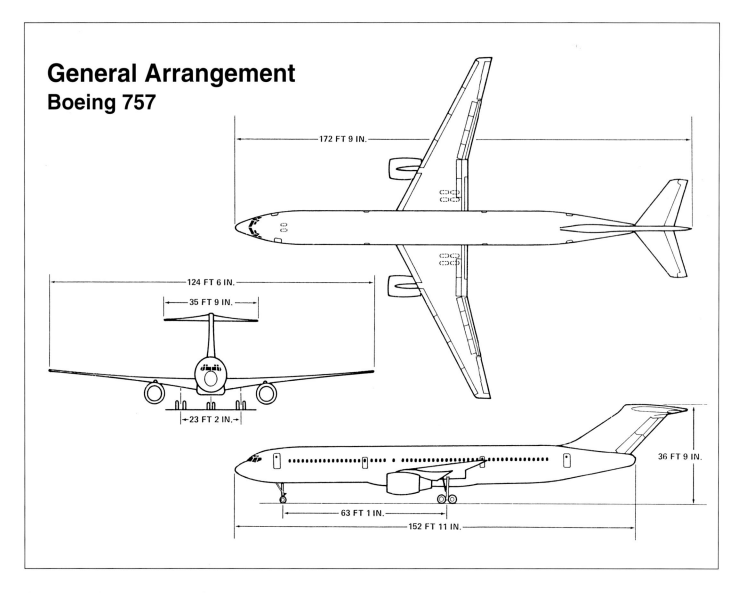

ABOVE: By 1978 the overall shape of today's 757 had emerged in the late 7N7/early 757 configuration. The aircraft still retained the T-tail, but within 12 months this feature would be deleted. *The Aviation Picture Library*

domestic market, having grown from a 120-seater, to a 160 to 180 seater. The change in fuselage section was influenced by the airlines, particularly United, who reacted against the 1950s-style standard narrow-body six-abreast cabin width. It was felt that greater width was needed to improve passenger comfort, and it was believed that there would be difficulties in stretching a long thin fuselage beyond 180-seat capacity, restricting development potential. The initial Boeing proposal was to offer a 6in increase in cabin width, but it was found that a wider fuselage would provide more flexibility in layouts.

This larger 7N7 was becoming a close challenge to the 7X7 in the market place, with little to choose between them. It had become a completely new type, moving away from the earlier derivative studies. This would increase its development costs and take up greater design and engineering resource — possibly delaying the launch of the wide body 7X7.

The seven-abreast configuration was not attractive to the airlines. It added one line of seats at the expense of a second aisle, and offered little passenger appeal over the eight abreast twin-aisle Airbus A300. The diameter was also unable to accept a pair of the standard LD3 cargo containers in the underfloor hold, which would restrict the additional valuable capability of carrying freight.

By early 1978 the configuration of the 7N7 had returned to the narrow-body configuration based on the 727 cross-section, flight deck and modified empennage, but fitted with a new wing. In this form it became known as the Boeing 757. With a 150- to 170-seat capacity (depending upon the arrangement) it corresponded closely to the 727-300B project abandoned three years previously. Power was planned to be from a pair of podded wing underslung cropped fan engines such as the GE CF6-32 or R-R RB.211-535s, with uprated P&W JT8D-209s or JT10D derivative as alternatives.

DEFINING THE 757

British Airways and Eastern Airlines emerged as the major potential customers for the 757, and they helped Boeing evolve the overall specification for the aircraft. The previously defined 150-seater was stretched by two seat rows to accommodate 162 passengers for Eastern, while British Airways planned to seat around 175 passengers in a high density single class layout. The unit price of the aircraft was set in the region of $14.75 million. The 757 was expected to have range and payload

EVOLUTION

ABOVE: The Model 757 designation was applied to the 7N7 design studies in early 1978, at the same time the Model 767 and the (initial) 777 were launched as firm projects. *The Aviation Picture Library*

advantages over the smaller, lower cost (but still competitive) DC-9-80 series, and at least one established DC-9 operator was showing interest in the projected 757. Providing the stretched DC-9-80 variants did not gain too much of an early hold on the market, the US domestic airlines were seen as the main initial sales targets for the aircraft.

As defined, the airframe was still based on old 727/737 structural technology, to keep costs down, but the airlines appeared to be prepared to pay up to a further $2 million per aircraft to make them more in keeping with the latest design standards and to give a longer operational life. A further fuselage stretch adding room for another three seat-rows, brought capacity up to 180 seats, and with it the designation 757-200. This would become the basic production configuration for more than 900 aircraft that followed — the shorter 757-100 was never built. By now the wide body 7X7 had become the 767. The seating capacity of the 757-200 was similar to the short fuselage 767-100, but much less than the longer 767-200. This helped to define the latter aircraft. Even using updated structural technology, the $18 million 757-200 would be cheaper to buy than an all-new 767-100 costing $22.5 million, and seat/mile costs would be considerably less than any comparable wide-bodied airliner. With deregulation in the USA bringing down fares and therefore reducing yields, the smaller aircraft offered the advantage of greater frequencies and the cost of

BOEING 757

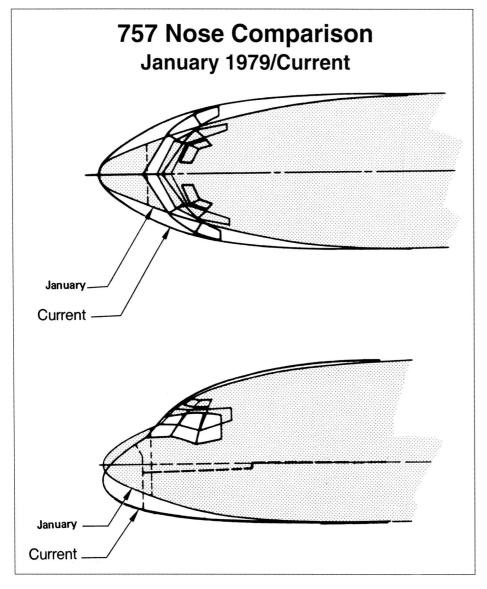

RIGHT: In 1979 the 757 was redesigned and restyled to become the aircraft we know today. Gone were the T-tail and the old-style nose, while the wing was further improved and the fuselage extended slightly. In this form the aircraft was now ready to be offered to the market. *The Aviation Picture Library*

LEFT: During 1979, when the 757 lost its T-tail and underwent other design refinements, one of the more subtle changes included was a redesign of the nose section. The original, narrow 727-style nose section was replaced with a wider, more rounded section, that was in keeping with the 767's. Both the 757 and 767 were then built around the same flightdeck design, which allowed a high degree of production-line and fleet commonality. *The Aviation Picture Library*

flying an extra aisle around the sky would be economically unattractive. The lower fares were expected to increase the 757-sized airliner market, possibly leading to increased traffic growth, and therefore creating a demand for the larger aircraft on the more popular routes.

ROLLS-ROYCE POWER

The Boeing 757 was initially offered with the R-R RB.211-535 turbofan as the favoured engine and the GE CF6-32 and P&W JT10D as options. The RB.211 was expected to provide a fuel burn of 25kg (56lb) per passenger over a 1,000km (500nm) sector at a gross weight of 87,180kg (192,200lb). Overall fuel consumption was about 15 percent better than the 727, and fuel burn per passenger was 30 percent improved. Take off distance of the 757 was planned to be around 1,800–3,000m (6,000–10,000ft) better than the 727. The four-wheel main undercarriage units would maintain a load classification number (LCN) below 50 allowing unrestricted operations from difficult airports such as La Guardia. The approach speed of 231km/h (125kt) gave a wet runway landing distance of less than 1,500m (5,000ft) and the 757 had an adequate wing area of 180sq m (1,950sq ft), to avoid the development problems experienced with the 727 and 737.

By May 1979 Boeing was committed to launch both the narrow-body 757 and the wide-body 767, defining the specification with the help of a number of prospective launch customers. The launch of both programmes at the same time would require a considerable investment by Boeing both financially and in engineering resource, but included advantages in lower costs by making the systems as similar as possible and sharing development time. The decisions on cabin widths had not been easy, as the narrow body worked well for smaller capacities, but had limited stretch capability, and a wide-body becomes too dumpy at the lower end of the scale bringing drag penalties.

The 757 was effectively launched by a simultaneous commitment for a total of 40 aircraft in August 1978, although Boeing did not confirm the formal launch of the programme until March 1979. The commitments were 19 aircraft for British Airways and 21 for Miami based Eastern Airlines, the R-R RB.211-535 turbofans being selected by both airlines to ensure a first flight before the end of 1981 to meet the planned in-service dates.

The 767 had been launched by an order from United in July 1978, who selected the larger capacity aircraft initially,

EVOLUTION

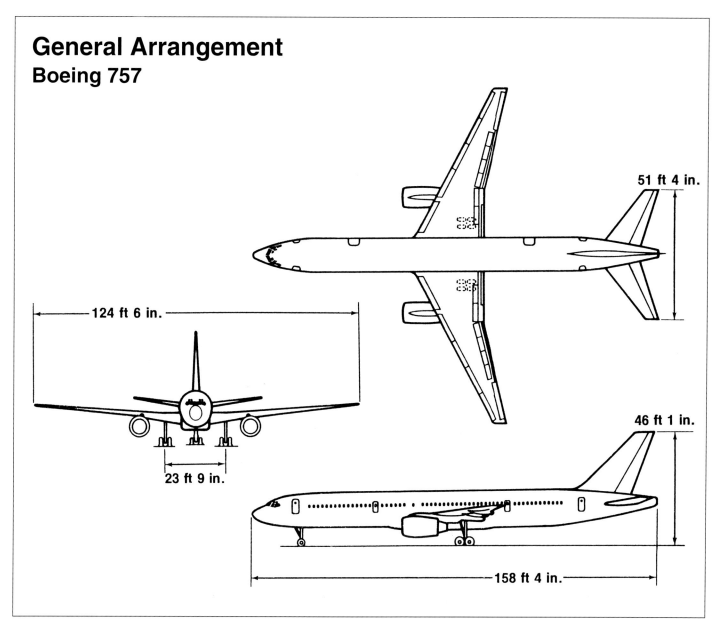

General Arrangement
Boeing 757

later placing major orders for the smaller 757. The modest gap between the launch of the two types helped ensure a more even distribution of funding and to ensure adequate engineering resources the 757 and 767 were allocated to different design teams and two manufacturing locations. The 767 assembly line was placed at the wide-body facility at Everett, with a new two bay assembly facility built next to the 747 plant. The 757 was built at the existing narrow-body airliner plant at Renton where the 707, 727 and 737s had been produced. This at least allowed a more modest investment in facilities for the 757.

PARALLELLING THE 767

The 757 design team proposed the same fuselage cross-section as used on the earlier jets but with more advanced materials where appropriate, and achieved improved performance and economy by designing an all new wing. This gained technology from the wing under development for the 767, but with the addition of more complex trailing edge flaps. Although the 757 was viewed as a smaller and lower risk programme, it was certainly not second best. The two design teams gained mutual benefit from as much systems commonality as possible, particularly on the flight deck and the avionics. As a result the 727-style V-shaped windscreen was replaced by a curved windscreen requiring a new tapered forward fuselage section blending the flight deck into the parallel section of the cabin. The overall length was reduced slightly, but passenger accommodation was increased by three seats. Although this gave the nose a slightly unusual shape, it allowed commonality of flight deck layout, which following certification, allowed cross flight crew qualification between the two types. Other areas of commonality were the rear fuselage mounted auxiliary power units (APUs), electrical power generation and air-conditioning equipment. Operating economics were key to the programme, and the excellent seat/mile costs of the 757 were a major commercial influence.

Although it was the smaller of the two new airliners, the 757 was still a major project. It was only about 0.6m (2ft) longer than the earlier 727-200, but had an extra 8.2m (26.8ft) of cabin accommodation, with the flexibility for reduction or stretching. The 757 was originally proposed with a 727-type

T-tail, but the low fuselage-mounted tailplane was adopted at the very last minute. Where practical, many components were shared with the 767, not only in the tail, but throughout the entire airframe. Savings were made in production, particularly with lower prices on the higher volumes of common equipment. The wings were planned with a sweepback of around 25° at quarter chord, with double-slotted trailing edge flaps and full span leading edge slats to provide good airfield operating characteristics.

The maiden flight of the 757 was due from Renton in February 1982 with up to five aircraft allocated to the flight development programme from nearby Boeing Field. The flight testing was expected to take seven months, totalling some 1,250 flying hours for an in-service date of August 1983 for the RB.211-535 powered aircraft. Should an airline select the 158kN (35,580lb) thrust GE CF6-32 turbofans, delivery would be some seven months later. Boeing was also planning a production rate of four to five aircraft per month, and expected to have 19 aircraft

EVOLUTION

ABOVE: This early artist's impression of the 757 in British Airways service shows an aircraft with echoes of the Airbus A320 in its lines. The final 757 shape was rather more angular and stalky. *The Aviation Picture Library*

completed by the time certification was achieved to allow early deliveries to BA and Eastern and any additional customers.

At the Paris Show in June 1979 Boeing showed models for the first time with the new nose shape incorporating the common flight deck with the 767. Having started out as a derivative of the 727, the 757 now only shared the fuselage cross-section. At that time two versions of the 757 were being offered, one with a gross weight of 99, 790m (220,000lb) carrying typically 196 passengers in a one class layout with 0.86m (34in) seat pitch over sectors of 3,700km (2,000nm), and the other with a gross weight of 104,326kg (230,000lb) over ranges of up to 4,600km (2,500nm). Both versions were offered with RB.211-535 engines, but GE power was still an option at that time. In addition to the firm order for 19 aircraft, BA added options on a further 18 aircraft to gain the benefits of the launch prices.

2 Design and Development

Following the decision to go ahead with the 757 programme, there was a period of about six months before the final contracts were signed with British Airways and Eastern. During this period Boeing was able to start the detailed design and plan the manufacturing phase which commenced on 23 March 1979. When the final specification was being established, many refinements were made along with two major configuration changes. One of these changes was to abandon the 727 flight-deck and adopt a common advanced technology cockpit common to the 767. The other change was to replace the swept T-tail derived from the 727, with a conventional swept fin and rudder and fuselage mounted swept tailplane. Despite the drag reductions predicted with the earlier layout, the low tail layout was able to meet the changed stability and control requirements of the longer aircraft with wing mounted engines. The low tail also gave the bonus of some 18ft overall reduction in length for the same cabin size, reducing centre of gravity (C of G) problems and improving ground manoeuvrability. The tail configuration change was only agreed days before contract signature by the launch airlines.

The design maximum take-off weight was initially offered at 99,790kg (220,000lb), with growth to 104,326kg (230,000lb) achieved by minor modifications to the existing fleet retrofitted after about a year in service, and an optional 240,000lb on new production aircraft. These increases in allowable gross weight would help in performance from hot-and-high or restricted airfields, since the wings and centre-section were already full of

BELOW: The Boeing 757 made extensive use of advanced composites, saving weight and avoiding corrosion. In addition to the external airframe parts, the cabin floors were also made from composites. *Boeing*

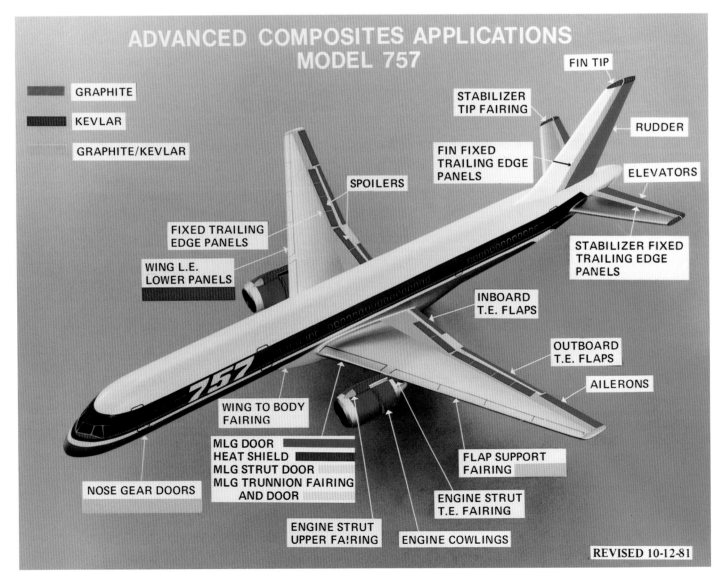

DESIGN AND DEVELOPMENT

ABOVE: In addition to being widely spread industrially in the USA, the 757 programme had a number of international partners and sub-contractors. *Boeing*

fuel, restricting any further capacity. Where there were no runway restrictions, the higher gross weight would allow 178 passengers to be flown 5,700km (3,080nm), instead of 3,980km (2,150nm) on typical US domestic services. For international operations, up to 223 passengers could be carried over 5,700km (2,520nm) as against 3,000km (1,610nm). The take-off field lengths required at sea level and a temperature of 29°C (84°F) was planned to be 2,100m (6,870ft), increasing to 2,320m (7,610ft) at 104,326kg (230,000lb), and 2,560m (8,410ft) at the maximum weight.

Some 70 percent of air travel consists of flights of less than two hours duration, and most short-haul travellers demand times that fit their schedules. It is therefore important for airlines to offer higher frequencies with smaller aircraft, rather than fewer trips with larger airliners, as is appropriate with long-haul operations. The 757 was being developed at a time of rising fuel prices and high inflation, making efficiency one of the most important factors in the specification. The highly efficient turbofan engines gave fuel savings of up to 20 percent over the earlier turbojets but attention to drag and aerodynamic refinements can contribute a further 10 percent. By retaining the narrow body six-abreast seat fuselage profile, seven percent was saved over the drag of a seven abreast twin-aisle aircraft. According to Boeing research, although passenger appeal is a critical factor, for trips under two hours a narrow body cabin is acceptable. The 757 was optimised over sectors of 500nm with an average flight duration of under two hours.

Typical customers, many of whom were 727 operators, were expected to have fuel economy as a priority and likely to be less interested in the carriage of large amounts of cargo offered by the wide-bodied airliners over longer ranges. The closest competitive threat was the McDonnell Douglas MD-80 series. With its lower purchase price, the MD-80 offered the deregulated and financially hard-pressed US domestic airlines an immediate, cost-effective answer to their requirements. The proposed Airbus A320 with its 150-seat capacity was seen as competition at the lower end of the capacity requirement, but not an immediate threat.

With both the 757 and 767 developed almost in parallel, though by different design teams at two different locations, it was logical to retain common design features wherever possible. This not only speeded development, but reduced costs, and

BOEING 757

LEFT: During static testing of the 757 airframe, the wing was bent upwards by over 139 in when it finally broke at about 112 percent of the ultimate design load. Results of these tests indicated an increase in gross weight was possible, to 240,000lb, without any strengthening of the structure. *Boeing*

made the family of airliners more attractive to airlines who planned to operate both types.

The 757 wing uses essentially the same aerofoil section as the 767, but it is smaller overall, and thinner at the root where it joins the fuselage. The wing section is a modern aft loaded design with a comparatively flat top, and a small under surface cusp near the trailing edge. The benefits include a delayed Mach drag rise, less wing sweep and a more efficient lighter and thicker wing structure, allowing more room for fuel and easier accommodation for the retracting main undercarriage. Although both the new Boeings were designed for a Mach 0.8 cruise the 757 only has a 25° sweepback, compared with 32.5° on the wide body 767. The reduced sweepback of the 757 was

DESIGN AND DEVELOPMENT

ABOVE: The 757 and 767 flight test programmes were running together. Although the 767 led the programme by five months, benefiting the common systems approach of the 757, the Boeing 757 led on the development of the two-pilot cockpit. *Boeing*

achieved without a large drag penalty due to a significant part of each flight involving either climbing or descending, on to the average shorter sectors. The short sectors also required the use of sophisticated high-lift devices on the wing, such as double-slotted flaps and full-span leading-edge slats without a drag penalty in the cruise. Due to the stiffer wing structure, the aileron reversal speed was raised avoiding the need for inboard ailerons. The larger wing span reduces induced drag, and the generous wing area is more able to cope with a fuselage stretch. The shorter fuselage achieved with the low set tailplane eliminates the possibility of ineffective pitch control at high nose-up atitudes. The two engines are located on underwing pylons to relieve bending moment and reducing structural weight.

Both the 757 and 767 airframes were designed for a life of 50,000 flights, the normal equivalent of 20 years of airline service, although account had to be taken of the larger number of sectors flown by the 757. With deliveries commencing at the end of 1982, many of the airframes will either have exceeded or be approaching the design life, but with careful structural monitoring the aircraft are expected to continue in service for many years with the minimum of modification or repair.

The Boeing systems philosophy calls for minimum crew actions in the event of a fault or failure, with the majority of systems transfers made automatically to alternative units. The engine indication and crew alerting system (EICAS) demonstrates this by making little provision for crew trouble shooting. This is justified by the high level of built-in automation, and the inability of the crew to make any adjustments once a fault has been identified. This philosophy implies comparatively simple systems and the two integral wing fuel tanks would have

BOEING 757

ABOVE: The first Boeing 757, N757A, was used for the more demanding aspects of the flight testing and remained the property of Boeing. The flight development programme was shared with four other 757s destined for delivery to BA and Eastern in the 10-month long certification programme. *Boeing*

been preferred, but to maintain a useful range, fuel is also carried in the wing centre section extending to between the two engine pylon locations. The fuel in the centre section is used first to maintain a good bending moment relief and put less stress on the structure.

Each engine drives a hydraulic system, and a third hydraulic system is driven by electric pumps using engine generated power, the pumps being the same as used on the 767. Also common to the 767 are the two 90kVA Sundstrand generators and electric signalling, rather than control cables, used to operate the spoilers.

The Rolls-Royce RB.211-535 cropped fan engine developing initially 166kN (37,400lb) thrust, was the lead engine for the 757 programme, as it was the only suitable turbofan certificated for

BELOW: Boeing 757 No 1 made 28 velocity-minimum unstick take-offs from Edwards Air Force Base during a week of flight testing. The rear fuselage was protected from damage by an oak skid, and the test proved that the aircraft would take-off safely without stalling, even if the tail was scraping the ground. *Boeing*

DESIGN AND DEVELOPMENT

ABOVE: 757 No.1 demonstrated that when water was ingested by the engines, there was no reduction in performance. Eight runs were made at Boeing Field in Seattle, seven with the nose wheels through the water, and the eighth with one main gear to check spray flow at the rear of the aircraft. The yellow paint on the aircraft is a water-soluble coating to highlight water impingement patterns. *Boeing*

the aircraft, and it was selected by both BA and Eastern. Following these launch orders were a couple of smaller orders from Aloha and Transbrasil, both specifying the 162kN (36,500lb) thrust GE CF6-32C1 turbofans. Although developing less power it compensated by being lighter, and the engine was only about a year beyond the RB.211 in its certification programme. But, with further substantial 757 orders from Delta and American who specified the P&W PW2037 turbofan developing 170kN (38,200lb) thrust, the Aloha order was cancelled, and Transbrasil adopted the PW2037, since without a substantial order, the GE engine was no longer economic to certificate. The 757 engine power has therefore been shared between Rolls-Royce and Pratt & Whitney. The PW2037 had emerged as a prospective power plant towards the end of 1980 and started as the newly developed JT10D.

With the price of fuel as a major factor in the operating costs of modern airliners, fuel economy is a major consideration, creating keen competition between the engine manufactures. Boeing's estimates for fuel burn with the 757 were expected to be as low as 27kg (59lb) per seat on the average 925km (500nm) US domestic one-class sector and 42–46kg (93–102lb) per seat on a 1,850km (1,000nm) flight. The improvements over the earlier generation aircraft were significant; the 737-200 using 21 to 25 percent more fuel per passenger, while the 727-200 used 36 to 39 percent more. The other advantages included considerably less noise and reduced emissions. Fuel-efficient turbofans also need to be fitted to a low drag airframe to maintain the lowest possible operating costs, and British Airways stated that the 757 featured the lowest seat/mile costs by a significant margin of any aircraft considered during the detailed and demanding evaluation.

Although the go-ahead for the 757 programme in March 1979 was nine months behind the larger 767, the systems commonality of the aircraft helped reduce the overall development programme to 35 months, as compared with 39 for the 767. As

the 767 engineering progressed, there was an orderly transfer of design personnel from the 767 programme to the 757, not only avoiding a short-term peak in skilled recourse, but also ensuring overall structural and systems commonality wherever possible. All the major milestones were achieved on, or ahead of schedule, with first metal cut on 10 December 1979 and final assembly commencing at Renton in January 1981.

The flight development programme was planned to take 1,284 hours, 65 fewer than the 767, with certification expected before the end of 1982 with deliveries in the following year. Rolls-Royce planned certification of the RB.211-535C by the end of March 1981, giving a comfortable margin for production engines to be available for the flight testing.

The maiden flight of the Boeing-owned first production

DESIGN AND DEVELOPMENT

ABOVE: To maintain the busy schedule, some of the 1,300-hour 757 flight testing was done at Palmdale and Edwards Air Force base in California. The Boeing-owned aircraft No. 1 carried the logos of the launch customers on the side of the fuselage. *Boeing*

aircraft, N757A, was from Renton on 19 February 1982, a week ahead of schedule, and after the 2hr 31min flight commanded by John Armstrong, a landing was made at the Boeing Flight Test Centre at Boeing Field close to Seattle. John Armstrong, the 757 Project Test Pilot reported that all the first flight goals were achieved, including checking all high lift devices and taking the aircraft to a maximum indicated air speed (IAS) of 460km/h (250kt). The aircraft was also simple to handle. This was the first flight of the Boeing new technology two-man cockpit, later to become common with the 767. In

ABOVE: Once tests had been completed with the Rolls-Royce RB211-535 turbofan, the Boeing 757 No.1 was re-engined with the Pratt & Whitney PW2037 engines in preparation for service with Delta, Singapore and Northwest. Warm weather tests were carried out at Tucson, Arizona as part of the two aircraft 340-flying hour programme leading to certification. *Boeing*

addition to the first 757, a further four customer aircraft were allocated to the flight testing alongside the busy 767 programme.

With certification planned 10 months after the first flight, Boeing commenced an intensive flight development programme, but the poor initial late winter weather typical of Washington State caused some delays, due to the flying being restricted to fair weather only avoiding undue risks. John Armstrong was accompanied on the maiden flight by Lew Wallick, the Boeing Commercial Flight Test Director, and having already flown the 767, he was optimistic of achieving a common type rating for the two airliners. Handling of the 757 was easy enough for it to be flown in formation with the 727 camera aircraft, and the early part of the programme concentrated on aerodynamics and handling, including high speed, low speed and manoeuvrability checks. Much of the testing was concentrated on the new flight deck avionics shared with the 767, but the 757 lead the way, as the 767 was still initially configured for three crew operation, later changed to two crew. The new flat screen CRT displays were easy to read in all conditions and easy to use. They were well lit for bright light conditions without glare at night, and did not suffer from parallax making readings more accurate than with the conventional analogue electromechanical instruments.

HRH The Duke of Edinburgh was the first non-Boeing pilot to fly the 757, during a visit to Seattle in April 1982. While Prince Philip occupied the left-hand seat, Lew Wallick sat in the right-hand seat, giving an idea of the high level of confidence in the new aircraft only two months into the flight testing. The second aircraft (N501EA) — the first for Eastern — joined the flight test programme on 28 March to concentrate on type certification. It was followed by N502EA on 29 April, N503EA on 4 June and N503EA on 2 July. All were used for various aspects of the development testing including systems, interiors, power plants, hot and cold weather trials and route proving. The first aircraft was used for the more demanding tests, such as maximum energy stop to test the brakes and Vr to ensure the aircraft would not stall on the ground during take-off if the tail was touching the runway. The PW2037 development was undertaken on the first aircraft for Delta N601DL, which was the 37th aircraft off the line, making its maiden flight on 1 June 1984.

By the middle of 1982 over 200 hours had been flown covering take-off, landing and cruise performance and noise monitoring all with encouraging results. Minimum unstick

DESIGN AND DEVELOPMENT

ABOVE: In April 1991 a CAAC 757-200 was used to demonstrate its performance from high altitude airports when a series of take-offs were made from the 11,621ft high airport at Lhasa, in Tibet. The airport at Lhasa is surrounded by mountain peaks stretching up to 5,000m (16,400ft) and above. *Boeing*

velocity tests, with the oak-protected tailskid scrapping the runway to avoid stalling on the ground were undertaken on the vast Edwards Air Force base in California. During these tests John Armstrong completed 28 tail dragging minimum speed take-offs during one week. By August the first aircraft for British Airways (G-BIKA) was in final assembly and made its maiden flight of 2hr and 23min on 28 October 1982, when the test schedule included initial flight control functions, engine testing and a fully automatic touch and go at Moses Lake Airport. This aircraft was allocated to the British CAA certification programme.

The overall flight development progressed so well that the fifth aircraft for Eastern was able to participate in a sales tour of South East Asia in August, including Japan Airlines. The same aircraft participated in the Farnborough Air Show the following month, as part of a demonstration tour around Europe. During October and November, the same 757 completed an intensive 16 nation sales demonstration tour of Europe, Africa, the Middle East, and North and South America. A total distance of 86,400km (46,660nm) were covered in 67 flights with no aircraft-related delays, helping to prove the systems' reliability when operated a long way from a regular base. During the European part of the tour, automatic landings were made in high cross-wind conditions at Stockholm and Copenhagen. The longest duration flight of the tour was from Miami to Buenos Aires, overflying Cuba, Panama, Ecuador and Bolivia. By the middle of November, when the aircraft returned from the tour, some 1,200 flying hours had been flown in the development programme and certification was only a month away. The fourth aircraft had already completed its development tasks, and was being refurbished for the delivery to Eastern Airlines.

As the results from the flight test data were analysed, the 757 was found to be three percent better on fuel burn than expected, have a 370km (200nm) increase in predicted range and the operating weight around 1,650kg (3,650lb) below the original 1979 specification. The aircraft was proving to be the most fuel-efficient airliner currently available for sectors of

ABOVE: The 757-200PF Package Freighter was launched by an order from UPS which now has a fleet of 75 of these aircraft. Boeing 757-24APF N402UP (c/n 23724, l/n 141) was the first to be delivered, on 17 September 1987. *Boeing*

RIGHT: The main floor of the 757PF can accommodate up to 14 standard igloo containers, with an additional bulk container at the rear. To provide the maximum volume the crew enter through a smaller access door located ahead of the normal passenger door. *Boeing*

around 1,850km (1,000nm). Meanwhile, the first aircraft were in final assembly for the British charter airlines, Monarch and Air Europe, the latter being produced to the same specification as the BA aircraft to allow inter-operability.

FAA certification for the 757 was awarded on 17 December 1982 with the first aircraft already delivered to Eastern for crew training and engineering familiarisation. CAA certification was achieved in January 1983, allowing BA to commence services from London to Belfast on 9 February, the first two aircraft having been delivered in January.

As commercial operations became established, the fuel economy and greater than expected endurance, allowed some airlines to begin transcontinental operations. This required the aircraft to comply with what was known as extended-range twin-jet operations, later to become known as ETOPS, where the aircraft could fly up to 180 minutes from an alternative airport in the unlikely event of an engine failure. Eastern was able to use the full payload range to fly the 7,320km (4,550 miles) between New York Newark Airport and either Los Angeles or San Francisco. Northwest were able to operate 757s between Washington Dulles and Seattle, while Delta flew 757s from their Atlanta hub to Seattle, as well as other destinations. All these long flights helped to prove the reliability of the aircraft, engines and systems in preparation for ETOPS (Extended-range Twin-engined OPerationS) on long transoceanic flights.

The major initial developments for the 757 were as a Combi or an all-cargo freighter. Following a survey in early 1985 of the overnight package market in the USA, and also the possible interest of the US Post Office, Boeing found that a possible market existed for 70 to 100 aircraft in the 757 class. By producing an aircraft specifically for freight carrying, rather than convert from a passenger configuration, Boeing were able to offer a more cost effective aircraft. The resulting increase in gross weight by 4.540kg (10,000lb) to 113,400kg (250,000lb) from the passenger version allowed some 1,110km (600nm) greater range. The 757PF can carry on the main deck up to 14 2.25 x 3.18m (88 x 125in) containers weighing 236kg (520lb) and with a volume of 12.3cu m (440cu ft) each. A total of 3,665kg (8,080lb) of cargo can be carried, the main deck capacity of 187cu m (6,680cu ft) being supplemented by a further lower deck bulk freight volume of 51cu m (1,830cu ft). The major differences between the 757PF and the standard passenger version are the deletion of cabin doors, windows and

DESIGN AND DEVELOPMENT

ABOVE: The Boeing-owned 757 No.1 N757A has been converted into a flying avionics laboratory for the F-22 Raptor advanced fighter programme. The aircraft has a replica of the F-22 nose grafted to the front pressure bulkhead for the initial tests, and later has been fitted with a sensor wing section above the 757 cockpit. *Boeing*

passenger systems. A new upward opening 3.4 x 2.2m (134 x 86in) cargo door is installed on the port side of the forward cabin between the leading edge of the wing and the crew compartment. A new 1.8m x 2.56m (22 x 55in) crew access hatch is fitted forward of the former No 1 passenger entry door, and a solid bulkhead with a sliding access door separates the crew area from the main deck cargo.

United Parcels placed the launch order for 20 757PFs on 31 December 1985 with options on a further 15, and the PW2040 turbofans were selected in the $850 million contract. The PW2040 engines have a thrust of 186kN (41,700lb), which is about five percent more than the engines fitted to the passenger version allowed economic operations from the UPS base at Louisville. The increased gross weight also gives the aircraft a transatlantic capability from the USA to Europe, independent of any ETOPS requirements, as the aircraft does not carry passengers. UPS did however equip the oceanic twins with the extended range package, similar to the higher redundancy modifications fitted to the 767-200ERs. The changes included modifications to the APU, cargo bay fire suppression, the addition of a hydraulic motor generator, and software changes to the EICAS. The UPS 757s also have provision for an auxiliary fuel tank to be installed in the lower aft cargo bay, adding a further 3,030–3,410l (800–900gal) of fuel. To save a further 363kg (800lb) of weight, carbon brakes were selected to replace the earlier steel units.

The first two 757PFs were delivered to UPS on 16 September 1987, in March 1989 when the operator confirmed ten of the 15 options and in November 1990 UPS ordered a further 20 757PFs and 41 options worth $1.7 billion with deliveries through to 2001. A surprise move in December 1991 was the selection of the Rolls-Royce RB.211-535 engines for the new firm order by UPS for the 757PFs for delivery between 1994 and 1997. The options were also to be powered by the RB.211-535 turbofans when confirmed, while the original batch remained P&W-powered. The UPS 757PF fleet has now reached a total of 75 aircraft, more than justifying the cost of developing this variant.

The first non-US operator of the 757PF was Ethiopian Airlines who took delivery of ET-AJS on 24 August 1990, followed by Zambian Airways, who leased one 757PF (9J-AFO) from Ansett Worldwide Aviation from October 1990. London Gatwick based Anglo Cargo leased a 757PF from Ansett from August 1991, but it was returned due to financial problems in January 1992, and taken over by Challenge Air Cargo based in Miami with a total of three 757PFs.

The Combi was less popular, as only one has been built as part of an order by Royal Nepal Airlines for two 757s on 17 February 1986. Like the bigger wide-body Combis, the 757-200 Combi has as mixed main-deck cargo and passenger capability. All the passenger windows and access doors are retained, but an upward-opening 3.4 x 2.2m (134 x 86in) cargo door is

DESIGN AND DEVELOPMENT

ABOVE: The first stretched 757-300 N757X made its maiden flight from Renton on 2 August 1998, landing at Boeing Field after the 2½ hour flight where the development programme proceeded. This aircraft was eventually refurbished for launch customer Condor (as D-ABOA) and handed over in June 1999. It joined Condor's initial batch of three 757-300s, delivered three months earlier. *Boeing*

installed in the port side of the cabin, similar to the 757PF. The Combi is capable of carrying two to four 108in standard containers on the main deck. With three containers the Combi can carry up to nine tons of cargo and between 123 and 148 passengers. The Royal Nepal 757 Combi (9N-ACB) is the high gross weight version at 11,400kg (25,000lb), powered by RB.211-535 turbofans and was delivered in September 1988.

In July 1992 the 757 powered by P&W PW2000 series of turbofans achieved 180min ETOPS from the FAA and it was celebrated by American Trans Air with a six-hour inaugural non-stop flight from Tucson Arizona to Honolulu in Hawaii.

The first major development the since the launch of the programme was the stretched fuselage 757-300, which was being defined by Boeing in March 1996, and the German charter airline Condor was favoured as a possible launch customer. The 757-300X was a simple stretch of the 757-200 with two plugs inserted in the cabin fore and aft of the wing, as well as some local structural strengthening of the wing, centre-section and undercarriage due to the increased weight. This increased capacity by 20 percent from 198 to 235 passengers in high density seating configuration, while maintaining the range close to the existing 5,550km (3,000nm). This was expected to reduce seat costs by about 10 percent, making the aircraft attractive to a number of the European charter operators. Engine power was increased modestly using derivatives of the existing RB211-535E4 and PW2043 turbofans.

The stretched 757-300 was formally launched at the Farnborough Air Show in September 1996 with an order from Condor for 12 RB.211-534 powered aircraft with 12 options, subject to approval from the airline Board and with deliveries to commence in 1999. The 23.3ft (7.1m) fuselage stretch would provide accommodation for up to 289 passengers, but Condor planned for a total of 252 seats to give a greater level of comfort for passengers who were prepared to pay a little more for their holiday flight. The underfloor cargo capacity was increased by 40 percent and the anticipated range was 6,430km (3,500nm).

The 757-300 was taken up on its maiden flight on 2 August 1998 by Leon Robert, the chief project pilot, assisted by senior project pilot, Jerry Whites. The take-off was made at a reduced weight of 84,600kg (186,400lb) to allow for the relatively short length of the Renton runway. The 757-300 (N757X, later D-ABOA) for Condor, was a little over speed at take-off to gain the necessary clearance. Although the V1 decision speed at 113 kt is close to the 757-200, the rotation and climb speeds are two to four knots faster at 229km/h (123kt) for Vr and 246km/h (133kt) for V2. A maximum altitude of 4,875m (16,000ft) and speed of 460km/h (250kt) were reached, and some general handling characteristics were investigated at different flap configurations. These handling test included initial slideslip, approach to the stall with initial buffet, and some flutter work with rudder type kicks to excite different modes and responses.

The new aircraft flew very much like the 757-200 and the first landing was made at the flight test centre at Boeing Field. The only snag recorded was the loss of the fin mounted static pressure trailing cone, preventing the collection of calibrated airspeed data. The actual indicated airspeed had to be estimated within +/- 9km/h (5kt), the cone being lost within 45 minutes of take-off, probably over the mountains of the Olympic Peninsula where most of the test flight was conducted.

The test programme continued with the assessment of initial handling qualities including more stalls, wind-up turns, full rudder sideslips and variations in the centre of gravity (C of G). By mid August with increased speeds allowed the programme included flutter clearances up to speeds of 740km/h (400kt/Mach 0.91). At the end of the month, the first aircraft visited Edwards Air Force Base for runway tests, including the effect of the longer fuselage. The aircraft was used for rejected take-off tests at weights of up to 124,400kg (274,000lb), 1,360kg (3,000lb) above the current maximum taxi weight. Speeds were taken up to 350km/h (190kt) before maximum braking was applied with fully worn brakes. The tests required the aircraft to come to a full stop before taxying off the runway and standing for five minutes before fire crews could spray the red-hot brake units with water. Both brake suppliers, Dunlop and BF Goodrich, passed these demanding tests.

A total of three 757-300s participated in the flight development programme, all destined for delivery to Condor. The programme called for some 725 flight hours with certification due in December 1998 and first deliveries the following month ready for the start of the 1999 summer holiday season.

The second aircraft, D-ABOB, joined the flight test programme on 4 September, followed by the third, D-ABOC, on 2 October, making up some of the time lost with the production delays. By mid November aircraft No.1 had flown 270 hours, aircraft No. 2 had reached 210 hours, and No. 3 had about five hours. The third aircraft had achieved relatively low hours due to being allocated initially for high energy radiation field (HERF) tests which involved checking the fuselage for HERF attenuation frequency and oscillations, an FAA and JAA requirement since the development of the original 757. The extended underfloor cargo compartments were fitted with a redesigned fire suppression system, requiring smoke penetration, detection and suppression tests. The new system meters out the halon gas rather than discharge in one go, when smoke is detected. The system is equipped with additional monitors to meet the European JAA requirements.

Meanwhile, as a result of the early flight test results, minor changes were made to the flying surfaces and control system. These changes included adding vortex generators to the leading edge of the outboard flap sections to improve roll characteristics and give even roll response. This modification was considered desirable, but was not urgent, and could be retrofitted to the 757-200s. Changes were also made to the scheduling of the control column deflection in pitch to make the -300 handling more like the -200.

During a short period on the ground in November, the production standard Honeywell air data inertial reference unit (ADIRU) was fitted to be used in final autoland trails following the main ones on aircraft No 2. Once all the correct data had

DESIGN AND DEVELOPMENT

LEFT: The first production 757-300 is pushed back for another test flight in the development programme. *Boeing*

been collected the yaw damper rudder ratio change stabiliser module was fitted (YSM) to improve ride quality by driving the rudder to damp out lateral vibrations. A spoiler rescheduling system to help reduce pitch-up in the event of a slow landing, is also programmed into YSM, modifying the spoiler deployment sequence if the pitch attitude exceeds 8° on approach. With all the spoilers deployed, the pitch angle would normally increase with the high probability of a tail strike, but the system automatically delays the deployment of all the spoilers reducing the chances of the rear fuselage hitting the ground. Although the longer 757-300 was expected to be more prone to tail strikes, in practice it was found not to be the case, even when involving Condor pilots in the flight testing. The final flutter tests in December reduced the aircraft empty weight by 340kg (750lb), due to the removal of precautionary 170kg (375lb) balance weights in the wing-tips.

The second test aircraft painted in Condor colours, built up flying hours at an unprecedented rate. In the first month it flew 120 hours, with 110 in the second month and 90 hours in the third. Normally testing at a fairly aggressive rate about 80 hours would be achieved, but the aircraft had a high reliability rate, and the flight test engineering team were able to make configuration changes with the minimum of delay, having gained experience with the 777 and next generation 737s. Tests included ground effects, autolanding and fuel consumption tests. The autoland test required the aircraft to chase the aftermath of Hurricane Mitch from the Eastern coast of the USA across the north Atlantic for side wind conditions at Keflavik in Iceland and Cork in Ireland for a 46km/h (25kt) buffeting headwind. The aircraft was also used for the testing of a nosegear water spray deflector, which was found to be unnecessary, much to the relief of Condor, who were concerned about compatibility of towbarless push-back system at Frankfurt. The aircraft was then used for the final series of certification testing, including avionics before being grounded in December to be refurbished ready for delivery.

The major testing for the third aircraft, which was also in Condor colours and fully furnished was the service ready demonstrations involving four days of route proving from Frankfurt in the early part of December. A number of operations were made without passengers to several of the principal Condor holiday destinations to prove the systems in service. Turn-round times were evaluated using airline staff in place of passengers, the 757-300 estimates being only four more minutes to load, and 2.5 minutes to unload compared with the 757-200. In practice it was found that due to improved procedures by Condor, and other features of the aircraft, turnaround times were reduced from 61 to 62 minutes on the 757-200s, to 55 to 58 minutes with the new aircraft.

In January 1999 the 757-300 was awarded the US FAA type certificate, the production certificate, 180-minute extended range twin-jet operations (ETOPS) approval and the European JAA validation. The clearances of the Rolls-Royce RB.211-535E4B-powered version of the aircraft followed a tight five-month flight development programme, the shortest period for any Boeing derivative since the 737-400 in 1988. The three test aircraft amassed 1,286 hours of ground testing and 912 flying hours over 356 sorties. The first three 757-300s were delivered to Condor in March, with four more due for delivery during 1999, and the remaining six during 2000. Other 757-300 customers included Icelandair and Arkia.

With sales of the 757 reducing, there was a need for Boeing to inject new life into the programme, as the 757-300 had been slow to attract more than 17 new orders. Studies were therefore focusing on the longer range 757-200X, with the same overall dimensions of the standard 757-200, but fitted with the structurally stronger wing and main undercarriage of the 757-300 and two additional auxiliary fuel tanks in the aft hold. The two tanks each holding up to 1,890l (500 US gal) would increase the range by between 555km/h (300nm) and 648km/h (350nm) to a total of 5,000nm, and the maximum take-off weight would be increased by 9,072kg (20,000lb) to 123,377kg (272,000lb).

The firm backlog of orders for the 757 had fallen below 100 aircraft by August 1999, with all remaining orders scheduled to be delivered by the end of 2000. In early 2000, a small flurry of 757-300 orders from TWA, American Trans Air and air JMC ensured 757 deliveries until 2002 at least, and prospects for the stretched 757 continue to improve. Boeing continues to work on its 757-200X studies with a group of six key scheduled and charter airlines, including Continental and Air 2000 — both long term customers of the standard 757.

3 Production

The 757 assembly line was set up at the narrow body plant at Renton to the east of Seattle, alongside the existing 737 line. Some 40 percent of the 757 was created by computer-aided design (CAD), following the lead established by the 767. The Boeing structural philosophy was to fabricate major assemblies, requiring more tooling and labour, compared with the Airbus emphasis on integrally machined parts.

Although the fuselage section was based on the earlier Boeing airliners, weight reduction was achieved by the use of improved aluminium alloys, and composite parts in non-load bearing structures. A 900kg (2,000lb) weight reduction was achieved giving an average annual fuel saving of up to 113,550l (30,000 US gallons), assuming 1,400 sorties over typical sectors of 1,850km (1,000nm). The new aluminium alloys offered between 5 and 13 percent improved strength, without any loss in fatigue properties, corrosion resistance or toughness. The increased strength was achieved by a tighter control on the amounts of copper and zinc introduced in the alloy, the most critical structural areas being wing skins, stringers and the lower spar booms. The use of 5,160kg (11,380lb) of improved aluminium saved 277kg (610lb).

BELOW: The 757's wings are built by Boeing and joined on the production line at Renton to the centre-section, provided by Textron Aerostructures, in Nashville. Here, the wings for the first aircraft are seen ready for final assembly. *Boeing*

PRODUCTION

ABOVE: The forward and rear cabin sections of the Boeing 757 No. 1 were joined to the centre-section on 18 and 19 September 1981, followed by the tailplane. *Boeing*

The two main composite materials used were carbon fibre reinforced plastic (CFRP) and Kevlar reinforced plastic. The CFRP applications included include elevators, rudder, ailerons, spoilers and engine cowlings. The majority of the access panels, undercarriage doors, and the wing-to-fuselage and flap track fairings were made from a Kevlar/CFRP hybrid. Kevlar reinforced plastic was also used for some access panels, engine pylon fairings and the tips of the fin and tailplane. A total of 1,515kg (3,340lb) of composites were used in the 757, giving a weight saving of 676kg (1,490lb). After a stringent weight saving programme during the three-year design and engineering programme, a further 680kg (1,500lb) was saved, allowing a significant increase in range capability which was particularly helpful for European operations allowing regular flights to the Middle East, and transatlantic charter operations.

Boeing was responsible for the production of nearly 50 percent of the parts for the first 200 aircraft, the remainder being supplied by mainly US-based subcontractors. The Boeing Military Aeroplane Division at Wichita, Kansas is responsible for the nose and cockpit section and the Vertol Division produces wing fixed leading edges. Wing skins, rudder,

BOEING 757

ABOVE: The production line at Renton never stops: while the first 757 was being assembled, the wing and cabin centre-section for the second aircraft was being fabricated close by. *Boeing*

RIGHT: Once full production of the 757 was established, the aircraft were lined up on both sides of the assembly building at Renton with the engines fitted at the final stage before painting. *Boeing*

elevators and access doors are produced at Renton where final assembly replaced the running down 727 line. Following the initial two deliveries at the end of 1982, production rate was planned to reach 2.5 per month by the end of 1983, with a total of 30 delivered during the year. The approximate cost of each 757 was expected to be $31–34 million at 1981 prices, depending upon the specification.

As with all new aircraft, the 757 was subjected to a comprehensive fatigue test programme simulating 100,000 flights

PRODUCTION

BOEING 757

PRODUCTION

ABOVE LEFT: The No. 1 Boeing 757 was rolled out from the Renton factory on 13 January 1982 in front of an audience of 12,000 guests and employees. Three more airframes are on the line to the right, with Boeing 737s on the left. *Boeing*

LEFT: The first 757-300 was rolled off the Renton production line on 31 May 1998 painted in the Boeing house colours for the flight development programme. On completion of the testing the aircraft was refurbished and delivered to Condor, the charter division of Lufthansa. *Boeing*

ABOVE: The stretched 757-300s are produced on the same line at Renton as the 757-200s. The first 757-300 is seen amongst the shorter 757-200s, with 737s in the background. *Boeing*

and 40 years of operation over a period of 14 months, allowing any structural problems to be corrected well ahead of the programme.

The manufacture of the first left wing main spar for the new stretched 757-300 started in September 1997 with the automated spar assembly tool (ASAT) being used for the first time on the 757. The ASAT is capable of automatic drilling and installation of more than 2,600 fasteners in the wing assembly process, reducing time and cost, as well as improving quality. Major assembly of sub-components of the 757-300 began at Renton in November 1997, the first aircraft sharing the line with the standard production 757-200s. The front and rear cabin sections were joined to the centre-section in March 1998, making the aircraft the longest single aisle twin-jet built at a total of $54.4 million.

The first 757-300 was completed and rolled out at Renton on 31 May, when Condor increased their order by one aircraft, but the aircraft was not ready for the maiden flight until 2 August. This was about a month later than planned, due to Boeing being busy overcoming the production delays caused by a high numbers of orders, delays in the delivery of systems and equipment and large volumes of out of sequence working, particularly with the new generation 737s also built at Renton.

Once completed, the 757s are normally painted at Renton. They are then flown to Boeing Field for production testing and customer acceptance. A total of 54 757s was delivered during 1998, and in 1999 production was running at five aircraft a month — though this was due to be scaled back. The peak production had been in 1992 when 102 757s were delivered and the rate reached 8.5 per month. By the end of 1999 Boeing had delivered 67 757s during the year and gained 18 orders, leaving a backlog of 81 aircraft.

4 Technical Specification

BOEING 757 TECHNICAL SPECIFICATION

VARIANT	757-200	757-200ER	757-200PF	757-200Combi	757-300
DIMENSIONS AND WEIGHTS					
Wing span (ft in/m):	124 10/38.05	124 10/38.05	124 10/ 38.05	124 10/ 38.05	124 10/ 38.05
Wing area (sq ft/sq m):	1,992/ 185.25	1,992/ 185.25	1,992/ 185.25	1,992/ 185.25	1,992/ 185.25
Wing sweep (degrees):	25	25	25	25	25
Length (ft in/m):	155 3/ 47.32	155 3/ 47.32	155 3/ 47.32	155 3/ 47.32	178 5/54.4
Height (ft in/m):	44 6/13.56	44 6/13.56	44 6/13.56	44 6/13.56	44 6/13.56
Cabin length (ft in/m):	118 5/36.09	118 5/36.09	118 5/36.09	118 5/36.09	141 7/43.17
Cabin width (ft in/m):	11 7/3.53	11 7/3.53	11 7/3.53	11 7/3.53	11 7/ 3.53
Max seating:	239	14/35/119	Zero	12/138	279
Cargo volume (cu ft/cu m):	1,790/ 50.7	1,830/	1,790/50.7	1,790/50.7	2,387/67.6
volume, deck (cu ft/cu m):	zero	zero	6,585	880	zero
Pallets:	zero	zero	15	3	zero
Fuel (US gal/litres):	11,253/42,592	11,253/42,592	11,253/42,592	11,253/42,592	11,490/43,490
Max t/o wt (lb/kg):	220,000/99,790	250,000/113,395	250,000/113,395	240,000/108,860	270,000/122,470
Landing wt (lb/kg):	198,000/89,810	198,000/89,810	210,000/ 95,255	203,000/92,079	223,531/101,605
Zero fuel wt (lb/kg):	184,000/83,460	184,000/83,460	200,000/ 90,720	190,000/86,182	209,561/95,255
PERFORMANCE					
Max range (miles/km):	4,603/ 7,408	4,474/ 7,200	4,603/ 7,408	4,010/ 6,455	
Cruising speed (Mach):	0.80	0.80	0.80	0.80	0.80
Approach speed (mph/km/h)	152/245	152/245	152/245	152/245	152/245
Ceiling (ft/m):	42,000/12801	42,000/12801	42,000/12801	42,000/12801	42,000/12801

ABOVE: As a demonstration of the reliability of the Rolls-Royce RB211-535 turbofans, this Icelandair 757-200 held the record of 31,000 hours for the longest time on the wing before removal for an overhaul. *Rolls-Royce*

TECHNICAL SPECIFICATION

ABOVE: The Rolls-Royce RB.211-535C was the lead turbofan on the Boeing 757 programme, specified by both launch customers, BA and Eastern. *Rolls-Royce*

BELOW: The RB211.535-E4 was a development of the -535C, but the engines were not interchangeable, as modifications to the pylons were required. *Rolls-Royce*

POWER PLANTS:

Although all three major jet engine manufacturers offered turbofans for the Boeing 757 programme, only Rolls-Royce and Pratt & Whitney were selected at an early stage, leaving General Electric out of the project.

Rolls-Royce developed the cropped-fan three-shaft derivative of the larger 187kN (42,000lb) thrust RB.211-22B fan engine used in the wide-body airliners, such as the TriStar and Boeing 747. Design of the new engine, known as the RB.211-535, started at Derby in February 1977 with test-bench running commencing in April 1979. The RB.211-535C had a high pressure module based on the -22B, a six-stage IP compressor without variable stator vanes and a scaled down version of the more advanced RB.211-524 fans. Fan airflow was reduced by 18 percent from the -22B, and core airflow was 12 percent lower. The engine was designed to run at moderate temperatures, pressures and velocities, resulting in low noise of operation and optimised for short-haul operations.

By February 1980 five test engines had completed some 700 hours of bench running, simulating around 1,500 flight cycles from start-up to shut-down. Due to a high level of commonality with the earlier engines, only minor problems were encountered, and these were easily rectified. The high pressure module is normally the area of the highest technical risk, but this was

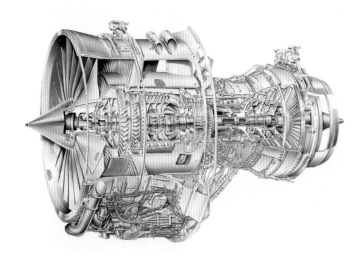

shared with the larger engine, allowing any -22B modifications to be easily embodied in the new -535. The high commonality of the engines also helped with spares and maintenance, particularly across mixed fleets. To maintain the best efficiency,

PW2037 TURBOFAN ENGINE

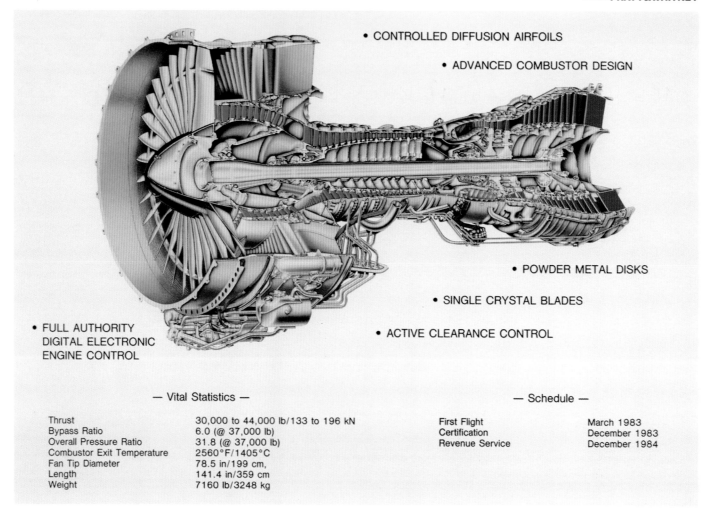

- CONTROLLED DIFFUSION AIRFOILS
- ADVANCED COMBUSTOR DESIGN
- POWDER METAL DISKS
- SINGLE CRYSTAL BLADES
- ACTIVE CLEARANCE CONTROL
- FULL AUTHORITY DIGITAL ELECTRONIC ENGINE CONTROL

— Vital Statistics —

Thrust	30,000 to 44,000 lb/133 to 196 kN
Bypass Ratio	6.0 (@ 37,000 lb)
Overall Pressure Ratio	31.8 (@ 37,000 lb)
Combustor Exit Temperature	2560°F/1405°C
Fan Tip Diameter	78.5 in/199 cm,
Length	141.4 in/359 cm
Weight	7160 lb/3248 kg

— Schedule —

First Flight	March 1983
Certification	December 1983
Revenue Service	December 1984

Rolls-Royce had responsibility for the complete power plant, and its installation to the pylon in co-operation with Boeing, including the basic engine, accessories and nacelle. To keep weight to a minimum, the nacelle made extensive use of composites.

To achieve certification by the Spring of 1981, Rolls-Royce undertook up to 3,000 hours of test-bed running, followed by a further 4,000 hours, including flight testing on the 757 flight development aircraft, ready for the planned service entry at the start of 1983. As part of the certification process, there was a simultaneous firing of eight birds into a running engine, to simulate a multiple bird strike, which the engine passed successfully. By the time the engine entered service on 1 January 1983, 36 engines had been delivered to the 757 production line. The intensive pre-service testing resulted in a highly reliable engine from the start of commercial operations. The RB.211-535C turbofan was initially rated at 166kN (37,400lb) maximum take-off thrust, which was sufficient power for normal temperate sea-level operations, and did not overload the engines when operating from hot and high airfields. The RB.211-535 was the lead engine on the 757, with the two launch customers selecting it for their fleets.

On 23 January 1982, 10 days after the formal roll-out of the first Boeing 757, the RB.211-535C engines were run for the

ABOVE: The 757-200 powered by the Pratt & Whitney PW2037 turbofan was launched into service through an order from Delta Air Lines. *P&W*

first time on the aircraft. The following day full power runs and trim test were carried out, followed on the next day by calibration checks, fuel flow measurements, acceleration responses and checks to the operation of the electronic control system. Once the combined ground testing and preparation of the airframe, engines and systems was completed, the aircraft was taken on its maiden flight on 19 February from Renton.

IMPROVED RB.211

Rolls-Royce was soon able to offer an improved version of the engine, designated the RB.211-535E4, with power increased to 178kN (40,100lb) maximum take-off thrust, and an eight per cent overall greater fuel efficiency in the cruise. Major changes included the use of wide chord fan blades to improve efficiency; a pressure ratio increase for the high pressure module; and a common exhaust nozzle for the fan and core streams. The higher efficiency core of the -535E4 had the pressure ratio increased to 27:1, as compared with 23:1 with the -535C engine, which contributed three percent to the improved fuel consumption. Since the 757 was originally optimised for short haul

TECHNICAL SPECIFICATION

ABOVE: The Boeing 757 flight deck was configured for a two-pilot operation and introduced for the first time (in a Boeing aircraft) eight CRT EFIS screens for flight and systems information. The 757 flight deck is identical to the Boeing 767, allowing crews to be qualified on both types. *Boeing*

operations, it was expected that about half the fuel would be used during the climb after take-off, the 11 percent savings with the E4 engine representing a major economy.

The RB.211-535E4 turbofan was awarded the type certificate on 30 November 1983 after running on the test-beds. It was first flown on the Boeing owned 757 No 1 in February 1984, at about the same time that Rolls-Royce and General Electric signed a partnership deal, giving GE a 20 percent share in each engine. The RB.211-535E4 engine was certificated at a thrust of 189kN (42,500lb) for the 757 at the end of 1988 and had entered revenue service in October 1984.

The RB.211-535C engines could not be modified to the new standard, requiring replacement if the benefits of the E4 engine were to be appreciated by the early operators. Only British Airways elected to continue with the -535C powered 757s as the Shuttle and European services did not really need the extra power, until finally specifying the E4 engines, which became standard fit for all operators who selected Rolls-Royce to power their aircraft. The other early airline customers had the engines changed for the improved version as they became available.

PRATT & WHITNEY POWER

The Pratt & Whitney PW2037 is a two shaft turbofan, previously known as the JT10D, and was launched on the 757 in December 1980 with an order for 60 aircraft by Delta. While both R-R and GE were offering derivative engines with lower costs and greater reliability, and possibly less development potential, the PW2000 series was a second generation high by-pass turbofan. Development of the PW2000 started in February 1972 with a 102kN (23,000lb) thrust engine, which ran for the first time on the test-bed in August 1974. Since the initial runs major developments were made with the engine, including the incorporation of advanced technology features. The development programme continued with the 142kN (32,000lb) JT10D-232 from January 1980, and in the middle of the year the engine was again scaled up for the 757. It was redesignated the PW2037, the last two digits denoting the thrust level in thousands of pounds. Initially, Rolls-Royce had a 34 percent share in the PW2037, but decided to withdraw in a changing market, and concentrate on its own product development. However, in July 1977 MTU in Germany and Fiat in Italy became collaborative partners with P&W, taking 11.2 and four percent respectively. In 1987 Volvo of Sweden became a four percent partner in the manufacturing programme.

With progressive development, the PW2037 has become a very fuel efficient, high technology turbofan developing 167kN (37,600lb) thrust. The technology advances used in the

ABOVE: The narrow body Boeing 757 cabin was fitted with traditional overhead bins and the business class cabin was typically five abreast. *Boeing*

PW2037 turbofan includes single crystal turbine blades, increased strength disc material, aerodynamically superior aerofoils and an electronic control system. The greater efficiency of the engine resulted in specific fuel consumption being reduced by 30 percent when compared with first generation turbofans, and configuration changes made the engine lighter, resulting in a thrust to weight ratio better than 5.5:1. Low noise features developed with the JT9D were included in the PW2037, including the use of a single-stage fan without inlet guide vanes, wide axial separation between the blades and vane rows, and a moderate fan tip speed. The engine was configured with a low-emission burner, and it was designed to be compatible with an acoustically treated nacelle to achieve noise levels to meet all predicted future requirements.

It ran for the first time on the test-bed in the production form in December 1981. A total of 11 development engines were involved in certification testing in some 5,500 hours running time. A further 5,000 hours were run following certification, and prior to entry into commercial service. To increase high compressor efficiency, the core rotational speeds were increased significantly, by driving a smaller compressor much faster. This gave the advantage of lighter weight and smaller frontal area, but the higher rim speeds increased the centrifugal loads, requiring strengthening of the disc to blade attachments. The modular construction allows easier maintenance by replacement of modules without removal of the engine from the wing, and most of the rotating parts can be inspected *in situ*.

Flight testing commenced with the PW2000 fitted to the Pratt & Whitney-owned Boeing 747 flying test-bed in February 1983, allowing the first six engines to be cleared for flight testing on the 757 by the end of the year. The Boeing-owned 757 No 1 was fitted with the initial set of PW2037 engines on 14 March 1984, with certification scheduled for October, and service entry with Delta at the end of the year.

ENGINE DEVELOPMENTS

The production turbofan develops 167kN (37,600lb) take-off thrust with a by-pass ratio of 5.8 and production soon built up to 20 to 30 engines per month. The engine was certificated to a power of 173.5kN (39,000lb) of thrust, with development growth to 200.25 (45,000lb). To ensure successful and reliable commercial operation, Pratt & Whitney continued to run a PW2037 on the test-bed on the Pacer Programme, endurance testing at least two years ahead of the airline fleet in terms of operating hours and cycles. A further development for the higher gross weight 757PF at 113,400kg (250,000lb) was the 185.6kN (41,700lb) thrust PW2040, which was selected by UPS for their initial batch of freighters. However, as reported elsewhere, UPS selected the RB.211-535E4 turbofans for the later aircraft in their fleet. Further developments of the PW2000 series included new HP turbine blades, new fan blades and new fan acoustic treatment, these improvements entering service for the first time in March 1994.

TECHNICAL SPECIFICATION

ABOVE: The 757 has conventional single all-speed ailerons fitted to each wing, and double slotted flaps are located inboard, either side of the engine exhaust. *Philip Birtles*

ABOVE: The Garrett GTCP331-200 auxiliary power unit (APU) is mounted in the rear fuselage. *Philip Birtles*

LEFT: The steerable nose wheel unit is fitted with two wheels and retracts forward. *Philip Birtles*

In addition to the main propulsion turbofans, a Garrett GTCP 331-200 auxiliary power unit (APU) is mounted in the extreme tail cone with an air inlet in the fin. The APU is a miniature jet engine providing power on the ground for the aircraft systems and starting for the engines. It can also be run in the air to provide emergency power for the systems in the unlikely event of an engine failure.

FLIGHT DECK

The 757 flight deck was designed from the start for a two-pilot operation, providing a spacious interior with 61cm (24in) greater width than the aircraft which it replaced. Amongst the improvements on the flight deck were the latest technology systems. All in-flight controls and displays are accessible and visible to both pilots, with unobstructed instrument panel visibility. Comfort was improved with lower noise levels, improved ventilation and temperature control, and better external vision. In addition to the normal two operating crew seats, there is also

ABOVE: The main undercarriage of the 757 features a pair of inward retracting units with four wheels on each. *Philip Birtles*

ABOVE: The 757 has a conventional empennage with a swept fin and rudder and a fixed tailplane with elevators. The air intake for the APU is located at the base of the fin. *Philip Birtles*

an additional observer seat, with provision for a second observer seat if required.

The new technology fully integrated electronic systems equipping the flight deck included the flight management system (FMS), electronic flight instrument system (EFIS), engine indication and crew alerting system (EICAS), automatic flight control system (AFCS) and a thrust management system. Digital electronics allowed extensive use of cathode ray-tube (CRT) displays, with a total of eight on the 757 flight deck. These new systems offer improved redundancy, reliability and maintainability, and the lower crew work load made operation by two pilots feasible.

COCKPIT SYSTEMS

The FMS offered fully integrated digital avionics with performance optimisation for best economy, VOR/DME automatic tuning and the flexibility to incorporate future enhancements. The integrated automatic flight management system controls the autopilot and engine thrust management. The features included thrust limit protection; optimum cost flight profiles for climb, cruise and descent; Cat IIIb autoland and the automatic recording of system faults allowing interrogation by maintenance engineers after each flight. The flight management computer systems (FMCS) are located on either side, and at the forward part of the central console between the pilots. It allows the crew to maintain performance management including optimum climb, cruise, descent and holding profiles; cost index selection; speed schedules and monitor engine-out performance. The FMCS assists with flight planning, including providing range/fuel data, take-off and landing information, advised top of descent point, and gives optimum altitude and steep climb points. The navigation/guidance calculations include point to point on a great circle route, altitude/speed profiles, guidance commands and the database stores aircraft/engine performance and route navigation information.

In front of each pilot is a pair of vertically mounted colour CRTs, the top one being the attitude director indicator (ADI) and the lower one the horizontal situation indicator (HSI). The ADI replaces the old artificial horizon, giving information on roll, pitch and slip. In addition ground speed is indicated and autopilot status. The HSI provides navigational data replacing the old moving map. It can be selected in the map mode showing track, distance to go with estimated time of arrival, aircraft heading and wind speed and direction. Additional modes which can be selected include the VOR or ILS. with weather indicated.

Located in the centre of the main instrument panel in front of the pilots is a pair of the EICAS colour CRTs which provide information selected by the crew, or what is relevant to the particular stage of operation. The top EICAS CRT shows warnings, cautions and advisory information and the engine status primary information. The lower CRT usually is dedicated to engine secondary information, systems status or maintenance display. The major advantages include centralised engine and alert displays, reduced pilot and maintenance crew workload, automatic and manual event recording, increased display and interface flexibility, reduced costs of ownership and the capability of future software development.

TECHNICAL SPECIFICATION

A TYPICAL TURNAROUND
Britannia Airways kindly arranged for an air-side apron pass to cover the turnaround of a 757 at the Luton operating base. The aircraft concerned was 757-204 G-BYAI (c/n 26967, l/n 522), which first flew from Renton on 26 January 1993 and was delivered to Britannia Airways on 1 March the same year. The aircraft was observed during April 1999 on the new apron at Luton, known as 'the six-pack' where there is space for six airliners by the new terminal building. *All photos by the author*

ABOVE: Britannia Airways Flight BY422B turns towards its stand, on arrival at Luton from Alicante.

LEFT: As the aircraft comes to a halt, chocks are placed by the nose-wheels.

BELOW: The sun-tanned passengers disembark from the middle door on to a cold and damp apron.

The crew alerting system provides automatic system monitoring with three levels of visual/aural/tactile alerts when required. The three levels of crew alert are red for immediate attention, amber for operational and green for information and non-essential data.

The automatic flight controls consist of the autopilot/flight director, the auxiliary controls and thrust management. The autopilot provides direction of the control wheel steering, vertical speed select and hold, altitude select and hold, IAS/Mach select and hold, heading select and hold, vertical and lateral steering, and approach, landing or go-around information. The auxiliary control functions include spoilers, yaw damping, rudder authority, stabiliser/Mach speed trim and elevator asymmetry. The thrust management computes the rating limits, maintains working temperatures with autothrottle functions and maintains IAS/Mach hold according to the phase of operation. The autopilot flight director system (AFDS) has two modes of operation, pitch and roll. The pitch mode controls take-off using the flight director, altitude select and hold, flight level change, vertical speed and navigation, the approach, go-around and flare and control wheel steering through the autopilot. The roll mode controls the selection and hold of heading, approach and go-around, ILS localiser beam, rollout control, lateral navigation and control wheel steering.

FLYING CONTROLS
The 757 has conventional flight controls consisting of single all-speed ailerons, single rudder and elevator control surfaces, flight spoilers, slats and double slotted flaps. The spoilers are also used for speed brake control. The primary flight controls

BOEING 757

ABOVE: A Servisair team begins to unload the baggage from the forward hold with access on the starboard side.

ABOVE: With passengers loading and unloading on the port side, all services are on the starboard side. The catering vehicle services the rear galley.

ABOVE: Refuelling is through a single point under the starboard wing.

ABOVE: In-flight catering is also serviced through the forward service door.

are fully powered by three independent hydraulic systems. Lateral control is provided by the pair of ailerons, assisted by the wing mounted spoilers on the upper surfaces. Directional control is through the rudder mounted on the vertical fin with a yaw damper to smooth out the ride. Longitudinal control is by a pair of elevators mounted on the trailing edge of the fixed tailplane. The high lift system consists of wing leading edge mounted slats, and double slotted wing trailing edge flaps.

STRUCTURAL DESIGN

The use of lightweight materials include a combination of improved aluminium alloys and advanced composites. This ensures a lightweight, durable, failsafe structure offering lowest cost of ownership. Additional features included corrosion protection and ease of maintainability. The new aluminium alloys offer increased strength, greater toughness, and improved corrosion and fatigue resistance. Corrosion protection is also achieved by improved bonding, sealing and surface finish. The advanced aluminium alloys are used in the manufacture of the wing upper and lower skins with stringers, front and rear spar webs lower spar chords, centre-section keel beam chords and rear spar web.

Advanced graphite, kevlar or hybrid composites are used in all the movable control surfaces, wing to fuselage fairing, engine cowling components and strut fairings, flap support fairings, nose wheel doors, and other areas both external and internal.

The cabin floors are reinforced fibreglass sheets with nomex core, both for lightness and corrosion protection. This results in 676kg (1,490lb) weight savings per aircraft. The substitution of carbon brakes for steel units saves a further 295kg (650lb).

ELECTRICAL POWER

Under normal operations, electrical power is generated by two engine driven 90kVA integrated drive generators (IDG), with an additional IDG driven by the APU. A single IDG is capable of supplying all the essential power, and a 24V DC battery is used for back-up power. The electrical system includes main

TECHNICAL SPECIFICATION

ABOVE: The toilets have to be serviced during the turnaround, with access under the forward and rear vestibules.

ABOVE: The passengers board for their holiday in Malta, with those who are disabled lifted by special vehicle to access the forward service door.

ABOVE: Everyone aboard, doors shut and engines are started ready for push-back.

ABOVE: The engineer severs his intercom connection with the flight deck once the tug and tow-bar has been removed.

and standby systems which are isolated, and are compatible with the triple-redundant Category III autoland capability.

HYDRAULIC POWER

The hydraulic system features three independent power systems using titanium tubing with welded end fittings. The central hydraulic system is dedicated to the primary flight controls only, and for ease of maintenance there is a single point filling access with ground check out requiring only electrical power. There is a power transfer unit for backup landing gear and flap/slat actuation, and a belly mounted, gravity lowered ram air turbine (RAT) provides flight control in an emergency.

FUEL SYSTEM

Fuel is carried in three integral sealed tanks in the wing struc-

ture. The centre-section tank holds 26,123l (6,901US gal), and each of the outer tanks have a capacity for 8,237l (2,176US gal) each. The joint of the tanks is adjacent to the engine pylons, with a dry bay by the leading edge, and surge tanks at the wing tips. There is a single point pressure refuelling panel in the leading edge of the starboard wing, outboard of the engine nacelle.

UNDERCARRIAGE

The main gear has a four-wheel truck mounted on the lower inner wing, and retracting sideways to be stowed in the wing root with a fairing door covering it after retraction. The steerable twin wheel nose unit retracts forward. An anti-skid system is fitted to the main wheel brake units.

5 IN SERVICE

The simultaneous selection by British Airways and Eastern Airlines of the 757 on 31 August 1978 launched the new airliner into production. BA had requirements for 19 aircraft with options on a further 18, and Miami based Eastern required 21 aircraft with options on 24 more. Both airlines chose the Rolls-Royce RB.211-535 turbofans following a long and hard sales campaign. The aircraft had accommodation for 164 mixed class passengers, or a maximum of 195 in one class. The British Airways contract, worth $300 million, was signed on 2 March 1979 for 19 aircraft with service entry planned in early 1993. Three months later options for 18 757s were confirmed. The BA aircraft were to be configured with 12 first-class seats and 174 economy class passengers, with the option of going to 200 passengers in all economy. As well as a full passenger load, up to 6,000kg (13,228lb) of cargo could be carried throughout the European network, and a number of the aircraft were allocated to BA Shuttle operations within the UK.

A contract worth $580 million with Eastern Airlines was signed on 23 March 1979, for 21 757s with options for 24 aircraft. An increase of four 757s was made in July 1980, with the number of options remaining unchanged. The programme could therefore be launched with total firm orders for 40 aircraft, but it was not until over a year later on 12 November 1980 that Delta Air Lines and Boeing made a joint announcement for the purchase by the airline of 60 757s.

The Delta order was covered by multiple contracts representing an overall investment of some $3 billion with deliveries spread from late 1984 through to 1990. The specification called for a two-class configuration, seating up to 190 passengers on US domestic sectors — and replacing DC-9 and 727s on medium- and short-haul services. Examples of the routes to be flown by the 757s included Memphis–Chicago, Seattle–Dallas/Fort Worth, Atlanta–Savannah, Chicago–Cincinnati, Atlanta–Jackson and Boston–Montreal. Delta was the first airline to select the Pratt & Whitney PW2037 turbofan in early 1981. Delta commenced scheduled services 757 in November 1984 from its main hub at Atlanta–Salt Lake City, later increasing the options from 10 to 20 aircraft with 28 already delivered. Following tests in September 1987, the Delta 757s were cleared to operate from the noise-sensitive John Wayne Airport in Orange County, California. Delta now has 102 757s in service, with a further 17 on order.

American Airlines was reported to have placed an order for 15 757s in January 1981, but this deal was not confirmed and the McDonnell Douglas MD-80 took an order for 100 aircraft instead. However, American did select the 757 at a later date. Other frustrated orders included three Rolls-Royce-powered 757s for Air Florida in August 1981, and two aircraft for Air Malta in September to replace the Boeing 720.

UK CHARTER OPERATIONS

The next positive order was for an initial batch of two 757s for Luton-based charter operator Monarch Airlines, placed on 19 February 1981. They were powered by the improved RB.211-535E4 turbofans to replace Boeing 720s. The aircraft are used on European and transatlantic holiday charter flights, the total fleet now consisting of seven aircraft.

Boeing achieved US-type certification for the 757 on 17 December 1982, about one month ahead of schedule, followed by UK CAA certification in mid January 1983. This allowed Boeing to bring the initial deliveries forward and the first,

BELOW: British Airways 757-236 G-BIKA (c/n 22172, l/n 9) made its maiden flight on 25 October 1982 and participated in the certification programme for the British C of A. Following refurbishment, it was delivered to BA on 28 March 1983. It is seen in the 'World' livery on approach to Heathrow in September 1999, however, it will soon be on the conversion programme to become a freighter for DHL. *Philip Birtles*

IN SERVICE

ABOVE: Eastern Airlines signed a contract for 21 757s on 23 March 1979, sharing the launch of the type with BA. Both airlines selected the Rolls-Royce RB211-535C turbofan. Boeing 757-225 N501EA (c/n 22191, l/n 2) made its maiden flight on 28 March 1982, joining the Boeing owned 757 No. 1 in the flight development programme. Following certification and refurbishment it was delivered on 18 August 1983 and since December 1994 has been in service with NASA as N557NA. *Boeing*

N506EA, was delivered to Eastern on 22 December. This allowed daily services to start on 1 January, on the Atlanta–Tampa and Atlanta–Miami services, carrying 148 passengers on the inaugural flight. With the delivery of further 757s, from 16 January, services were added to other US cities and Nassau. Eastern claimed that the 757s were the only aircraft which could fly a full load of 185 passengers 700 miles using 40l (10.4 US gal) of fuel per passenger. The overall fuel economy of the 757 was 50 percent better than the earlier generation jet airliners which it was replacing.

British Airways took delivery of its first 757, G-BIKB on 25 January 1983, followed by G-BIKC on 31 January. They were ready for services to start on the Heathrow–Belfast Shuttle on 9 February, beginning the replacement of Trident 3s on the Shuttle and other routes. The Glasgow Shuttle was added later the same month, and with three more aircraft delivered in March, the Shuttles to Manchester and Edinburgh were also taken over by 757s. The 757s were configured in a two-class Business and Economy layout, with 189 seats. As more aircraft were delivered, from the summer of 1983 the 757s were introduced on European routes including to Rome, Milan, Paris and Copenhagen, expanding to Athens, Nice, Amsterdam and Frankfurt in October, followed by Geneva and Zurich in January 1984. British Airways has continued to add to its 757 fleet to a total of 53 aircraft, specifying the RB.211-535E4 engines for the first time in 1988.

Finnair first showed interest in acquiring 757s in March 1981 due to an apparently attractive offer from Pratt & Whitney, but instead chose MD-80s (757s would later to be ordered). The Costa Rican-based airline, Lacsa also failed to confirm an order for two Rolls-Royce-powered 757s in April 1992.

Gatwick-based charter operator Air Europe placed an order for two RB.211-535C-powered 757s to the same specification

BELOW: Delta Air Lines launched the Pratt & Whitney PW2037 turbofan on the 757 when the initial order was placed in November 1980. Delta 757-232 (c/n 22814, l/n 61) was delivered on 14 May 1985 and is seen on approach to Orlando in August 1988. *Philip Birtles*

BOEING 757

ABOVE: Monarch Airlines of Luton was the first to order 757s for holiday charter operations. Monarch 757-2T7 G-MONB (c/n 22780, l/n 15) was not only the first 757 to be delivered to the airline on 22 March 1983, but also the first to be powered by RB.211-535E4 turbofans. *Boeing*

as British Airways, allowing the airline to achieve early deliveries at an advantageous price. These aircraft were later upgraded with the improved RB.211-535E4 engines. The two 757s were part of the BA order and, as part of the deal, the two airlines would swap-lease both 737s and 757s, reducing BA's capital needs by $40 million over two years. BA supplied the flight crews, simulator facilities, technical support and spares. The the first aircraft, G-BKRM, was delivered to London Gatwick on 6 April 1983 and entered service on 23 April. While waiting for the second aircraft, Air Europe leased an aircraft from BA to cover the busy summer season with an annual planned utilisation of 3,600 hours for each 757. Air Europe progressively added to their orders, with one 757 in May 1984, five in mid 1987 and 22 more in April 1988 subject to finance, to be delivered over a period of five years.

Monarch Airlines took delivery of its first 757 at Luton on 21 March 1983 with operations commencing on 26 March. The aircraft was configured with 228-seat 73–76cm (29–30in) pitch all tourist layout for holiday charter flights.

EXCELLENT FIGURES

Following three months of service in April 1983, the 757s with BA and Eastern had demonstrated good fuel efficiency and high reliability. By 28 March the nine aircraft in service had flown 2,190 hours in 1,330 revenue flights, achieving a 96.3 percent despatch reliability on departures within 15 minutes of scheduled. The first 757 delivered to Eastern had flown 585 hours by the end of March on 316 revenue operations and 10 of the 123 aircraft on order had been delivered — four each to BA and Eastern, and one each to Air Europe and Monarch.

After just over seven months in operation, in September 1983 the BA 757s were certificated for Cat 3B autoland, the minima being 150m runway visual range (RVR) and a 14ft decision height. Work continued towards Cat 3C which is theoretically zero-zero RVR and decision height, but in practice a visibility of 75m is the minimum maintained for normal operations to allow the crew to taxi the aircraft visually.

Singapore Airlines ordered four Pratt & Whitney-powered 757s on 31 May 1983 as well as six wide-body Airbus A310s, the first Asian order for both types. The intention of the airline was to compare the performance of the two types to allow a selection of a future type in larger numbers for the regional routes. The first 757 was delivered to Changi on 12 November 1984, and was introduced into service on 1 December with the A310. The 757s were used to serve Kuala Lumpur, Jakarta, Medan, Kuantan and Penang. SIA has a policy of maintaining a young fleet, and as the new 757s and A310s were delivered, they were replacing 10 Airbus A300s which had been in service for four years, giving an average fleet age of 27 months. SIA later decided to standardise on the A310s for the continental routes and the 757s were returned to ILFC who leased them to American Transair from November 1989.

On 25 August 1983, the new Munich-based charter airline, Lufttransport-Sud placed their initial order for two RB.211-535C-powered 757s with an option on a third aircraft, which was later confirmed. The value of the order was $85 million and the aircraft were used for flights from Southern Germany to holiday resorts in the Mediterranean and North Africa. The first 757 was delivered on 25 May 1984 ready for services to start on 7 June. At the end of 1987 the name of the airline was changed to LTU Sud and the fourth 757 to be delivered was the first to feature the new livery, with three more to follow. LTU Sud is part of the German LTU charter group which includes LTU, and operates a total of 12 757s. LTU also holds

IN SERVICE

ABOVE: Eastern took delivery of its first 757 on 22 December 1982 with commercial services starting on 1 January 1983. Eastern 757-225 N519EA (c/n 22209, l/n 40) was delivered on 21 November 1984 and is seen holding for take-off at Orlando in August 1988. It is now in service with American West as N915AW. *Philip Birtles*

a 25 percent share in the Spanish charter airline LTE with a fleet of three 757s.

After nine months in service 23 757s had been delivered to the first four airlines, flying a total of 25,729 hours during 15,260 revenue flights. Despatch reliability had reached an average of over 97 percent, with Monarch reporting a cumulative 98.6 percent with a fleet of four aircraft, and achieving 100 percent over a one-week period. The average daily utilisation of each 757 with Monarch was 7.2 flying hours, and one of the aircraft was the highest time 757 with 2,046 flying hours. One of BA's 757s led on landings with a total of 1,343.

NEW ORDER IMPETUS

After a three year lull in major sales for the 757, Northwest placed an order on 29 November 1983 for 20 Pratt & Whitney-powered aircraft, configured with 185 seats in a first class/tourist layout. The 757s were planned to supplement the existing fleet and to allow domestic route expansion. This order arrived in time to maintain a steady production rate, avoiding the need for an expensive reduction. A further order was placed in October 1985 by Northwest for 10 757s to replace Boeing 727s on transcontinental services, with deliveries between 1987 and 1989. By this time Pratt & Whitney had received orders for some 250 PW2037 turbofans for Delta, Singapore and Northwest.

The first 757 for Northwest, N501US, was delivered on 28 February 1985 with services commencing from Minneapolis St Paul on 15 March. Deliveries of the initial order being completed in 1989. In January 1988 three more P&W 757s were ordered with the cabins configured for 184 passengers. In 1986 Northwest acquired Republic to provide a domestic feed for the international operations. The Rolls-Royce-powered Republic 757s were returned to Boeing and exchanged for

BELOW: British Airways took delivery of its first two 757s in January 1983 for services on the airline's inter-city Shuttle. Enough aircraft had been delivered by March for Manchester to be added to the 757 Shuttle schedule. BA 757-236 G-BIKK (c/n 22182, l/n 30) was delivered on 1 February 1984 and is seen here landing at Manchester in January 1989. *Philip Birtles*

ABOVE: Gatwick based Air Europe placed an initial order for two 757-236s to the same specification as BA stipulated on 2 July 1982. The first aircraft, 757-236 G-BKRM (c/n 22176, l/n 14) was handed over on 30 March 1983 and arrived at London Gatwick on 6 April. It is seen here turning on to the stand at Manchester in October 1987 and is now in operation with Star Air Tours as N261PW. *Philip Birtles*

P&W-powered aircraft to maintain commonality across the fleet. In October 1989, Northwest further increased their 757 orders by 40 aircraft for delivery between 1993 and 1998, plus options on 40 more.

The reliability of the 757s continued to improve and with 35 RB.211-powered aircraft delivered in the first 18 months, despatch reliability had reached 97.8 percent. LTS recorded 100 per cent over a short period. Monarch was achieving 98.8 percent with the three aircraft delivered to date and Air Europe maintained the best sustained reliability of all — 99 percent on the highest daily utilisation of 10 hours per day. LTS followed with 9.5 hours a day. During the first 18 months the total fleet had flown 83,889 hours during 50,659 revenue operations. It was estimated that six million passengers had been carried over some 36 million miles.

The first European-based RB.211-535E4-powered 757 was delivered to Monarch at Luton in March 1985, the three earlier aircraft being later retrofitted with the improved turbofans. A further 757 was leased by Monarch from ILFC in May 1987 and two more were ordered in April 1988 bringing the total fleet to seven aircraft.

At the Paris Air Show on 30 May 1985 Boeing announced the first order for three aircraft, from Royal Brunei, for the extended range 757-200ER worth $175 million. The RB.211-535E4-powered aircraft were capable of flying from Brunei to Europe with one stop. The aircraft were configured in a spacious three-class 142-seat layout with 16 first class, 30 business class, and the remainder tourist class. The cabins were furnished with deep pile carpet, leather covered seats in the first class and gold-plated fittings in the roomy, well-appointed toilets. The first 757 (V8-RBA) was delivered to Royal Brunei on 6 May with revenue services starting soon after. The third aircraft was used for a while by the Sultan of Brunei and was transferred to the Kazakhstan Government in November 1995.

Boeing had hoped to sell 757s to Indian Airlines and had built the aircraft in anticipation of the order, but lost the deal at a late stage to Airbus Industrie. The Republic order for six Rolls-Royce RB.211-535E4-powered 757s, with six options, on 1 October 1985 was therefore welcome and ensured early deliveries. This was the second US airline to select the Rolls-Royce turbofans, the initial requirement being for 15 engines worth $84 million, and the choice was made due to superior reliability and lower cost of ownership. The first aircraft, N602RC was delivered on 6 December 1985, followed by two more before the end of the year ready for the busy festive season. After the take-over by Northwest all six Rolls-Royce-powered Republic 757s were acquired by American West through Boeing.

FIRST AFRICAN ORDER

Royal Air Maroc ordered two PW2037-powered standard-range 757s on 5 February 1986 — the first to be specified by an African/Middle East operator. The P&W turbofans gave a maximum gross take-off weight of 240,000lb and the cabin was configured with 20 first-class seats and 164 tourist seats. The first 757 for RAM (CN-RMT) was handed over on 15 July 1986 and set a new distance record for the type when it flew the 5,653 miles from Seattle–Casablanca non-stop. The second 757 (CN-RMZ) was delivered on 7 August and the aircraft replaced Boeing 707s on Middle East services, Casablanca–Jeddah being flown non-stop.

On 18 June 1986 Air 2000, the new Manchester-based charter operator confirmed an order for two 757s, one on lease from ILFC, and the other from Chemco Financial Services of New York. Rolls-Royce RB.211535E4 engines were selected for this $60 million order and the cabins were configured with 228 tourist class seats. The first aircraft, G-OOOA was handed over on 1 April 1987, arriving at Manchester two days later, followed by G-OOOB before the end of the month. The first

IN SERVICE

ABOVE: Singapore Airlines ordered four 757s in May 1983 powered by PW2037 turbofans for regional services commencing on 1 December 1984. The first 757-212 9V-SGK (c/n 23125, l/n 44) was leased from ILFC and delivered on 12 November. The aircraft later went into service with American Trans Air as N751AT and was bought by Delta in November 1996. *Boeing*

revenue flight was on 11 April, eventually covering some 15 Mediterranean destinations with an average 16 hours per day utilisation. To cope with the expanding network, Air 2000 ordered two more Etops capable 757s in November 1987 for delivery the following May to be used on the winter long-haul charter flights to Florida, Mexico, the Caribbean and Asia with one stop. Although the airline has the administration at Manchester, London Gatwick is a major operating base, with other flights from Glasgow. The Air 2000 fleet now includes 13 757s, which are operated seasonally and five more 757s with Canada 3000 which was formed in early 1989 to operate charter flights from Toronto to Florida, the Caribbean and Mexico.

El Al ordered three 757s in October 1986, subject to Israeli Government approval, which was confirmed a year later. Although El Al had traditionally used Pratt & Whitney engines, following lengthy discussions with current operators of the 757, the RB.211-535E4 turbofans were selected due to the proven greater reliability and efficiency. The cabin was equipped with a total of 191 seats in a two class business and economy configuration and the first aircraft, 4X-EBL was delivered on 25 November 1987, with a second arriving the following month. The 757s replaced 707s on services to Europe.

Sales had now reached only a rather disappointing 193 757s over an eight year period, even though the aircraft had built up an excellent reputation in service. However, it was not long before a number of significant orders were placed, in many cases by airlines who had developed routes with the older 737s and 727s, and required the capacity provided by the 757.

In early 1987 Phoenix-based American West ordered three RB.211-535E4-powered 757s with options on three more, which were soon confirmed. In December 1988 10 more were ordered with options on a further 15. The first six aircraft were the ex-Republic 757s and deliveries started with N901AW on 12 June 1987. Air Holland ordered three 757s in July 1987 for use on transatlantic charter flights from the spring of 1988, the first delivery being PH-AHE on 9 March. The airline experienced operational difficulties and ceased flying, but retained the

BELOW: The German charter operator LTU placed its initial order for two 757s in August 1983, and the first was delivered on 1 October 1987. A second aircraft, 757-2G5 D-AMUW (c/n 23929, l/n 153) was delivered to LTU Sud on 18 November 1987 and is now in service with the parent, LTU. *Boeing*

BOEING 757

ABOVE: In June 1986 Manchester-based Air 2000 placed an initial order for two 757s, later acquiring further aircraft some of which were from the redundant Eastern fleet. Boeing 757-23A G-OOOH (c/n 24293, l/n 220) was delivered new on 6 April 1989 and transferred to Canada 3000 as C-FOOH on 1 June 1995.
Philip Birtles

LEFT: On 25 May 1987 Air Holland leased three 757s from AWAS, later sub-chartering the aircraft to Britannia Airways. Boeing 757-23A PH-AHK (c/n 24291, l/n 215) was delivered on 2 March 1989 and is seen at London Gatwick.
Bruce Malcolm

aircraft for charter to third parties, some of the fleet being used by Britannia Airways during the 1991 holiday season, pending delivery of the first of their own aircraft the next year.

757 IN CHINA

In October 1987 CAAC placed the first of what was to become substantial orders for the rapidly expanding Chinese airlines. This initial order was for three RB.211-535E4-powered 757s, with an order for three more a year later. When commercial aviation was re-organised in China a number of new airlines were formed to serve the regions in the country, and the CAAC aircraft were allocated to China Southern. The first 757, B-2801, was delivered on 22 September 1987. Another new carrier, Shanghai Airlines, ordered three PW2037-powered 757s in December 1988, the first aircraft, B-2808 arriving on 8 August 1989, allocated initially to domestic services including to Beijing, Guanghou and Xian. Towards the end of 1991, Shanghai ordered a further five 757s. Although CAAC ceased to be an operator of aircraft, its role changed to co-ordinating orders and operations. It added 13 options in mid 1990 as part of an order involving a number of airliners, and converted them to firm orders in August 1991. Boeing 757s are now in service with China Southern, China Southwest, China Xinjiang, Shanghai Airlines and Xiamen.

With the improvement in sales and greater prospects, Boeing increased the production rate to four 757s per month. Orders had grown to 224 aircraft, with 137 delivered. Ansett ordered six 757s with six options, which were confirmed in July 1988, and followed up with 16 RB.211-535E4-powered 757s in November 1988, plus 10 more in April 1990 — all as part of their lease portfolio. Caledonian placed an initial order worth $100 million for two 757s with six options in early 1988, to the same basic specification as the BA aircraft, but-powered by RB.211-535E4 turbofans. London Gatwick-based Caledonian had previously been British Airtours, but with the take over of British Caledonian Airways by BA, Caledonian became the holiday charter airline of British Airways until sold to an independent travel company. At the end of the 1999 season, Caledonian was merged with Flying Colours to become jmc AIR with headquarters at Manchester, including a fleet of 12 757s. In May 1988 ILFC added orders for nine 757s for leasing.

IN SERVICE

ABOVE: El Al ordered three 757s on 1 October 1986. The first 757-258 4X-EBL (c/n 23917, l/n 152) was delivered on 25 November 1987 and is seen on a test flight prior to delivery. *Boeing*

At the end of May, Boeing received new commitments for 160 757s worth $6.6 billion. These orders were for 50 RB.211-535E4-powered aircraft for American Airlines, with options on another 50, and 30 orders with 30 options for PW2000-powered 757s for United. The American Airlines 757s were scheduled for delivery between 1989 and 1993 to replace noisy 727s and some 737s, and so the low noise of the Rolls-Royce engines proved critical. American ordered yet more 757-200s in mid 1991 and the total 757 fleet is now 102 aircraft. The United 757s were due for delivery between 1989 and 1991 to replace 29 DC-8s, bringing benefits of greater Boeing commonality and the efficiency of a two crew twin-jet airliner. In April 1989 United increased its orders for P&W-powered 757s with an additional 60 aircraft, plus a further 60 options. United now has 98 757s within the overall fleet. In September 1988 Delta added 50 options for 757s in addition to the previous orders, nine more at the end of 1990 and four more 757s in mid 1991. The 757 fleet totalled 102 aircraft at the end of 1999, with 17 more to be delivered.

The Canadian charter airline, Odyssey International, which was formed in December 1988, leased three 757-200ERs from ILFC to be used on regular charter operations from Toronto to Stansted, Bristol and Manchester. The first aircraft, C-FNBC, was delivered on 20 December 1988, but the airline ceased trading in June 1990 and the aircraft were returned to ILFC. Two were leased to Nationair who also were grounded in April 1993. At the end of April 1989 a £17 billion order was placed by a consortium of GPA and Rolls-Royce which included 50 757s. MEA had hoped to update its fleet with two 757s to replace 707s, but the political and economic situation was not then conducive to this plan.

Sterling Airways of Denmark placed its first order for 757s in 1989, specifying three RB.211-535E4-powered aircraft with options on three more, worth $150 million. The cabins were

ABOVE: CAAC of China placed an initial order for three 757s on 20 January 1987. These, and subsequent aircraft, were all allocated to local carriers when CAAC was split up into independent regional airlines. CAAC 757-232 (B-2806 c/n 24401, l/n 232) was allocated to China Southern and delivered on 28 August 1989. *Boeing*

configured with 219 tourist class seats and the first aircraft, OY-SHA, was delivered on 7 June 1991. Three initial aircraft were sub-leased from Air Holland, from October 1989. Unfortunately Sterling ceased operations in September 1993 and the aircraft were repossessed. In April 1990, Avensa of Venezuela acquired two of the ex-Air Europe 757s to replace 727s on its Caracas to Miami and New York services.

Icelandair took delivery of the first of eight 757-200ERs (TF-FIH) on 4 April 1990, replacing DC-8s on transatlantic services, such as Stockholm–Baltimore. The aircraft were

BOEING 757

ABOVE: Canadian charter operator Odyssey International leased three 757s from ILFC for regular transatlantic services. The first delivery was in December 1988, but the airline ceased operations in June 1990. Boeing 757-28A C-GTDL (c/n 24543, l/n 268) was delivered on 12 March 1990 and is seen here the following month at Manchester. It was transferred to American West as N911AW in May 1990, to Nationair as C-GNXU in October 1991 and to Air Transat in April 1993, with which it currently operates. *Philip Birtles*

configured with 22 business-class seats and 167 economy and, as the fleet has increased, the 757s have been used on the shorter routes, including to London. Two of the 757-200ERs are still to be delivered, and the airline has now also placed orders for two stretched 757-300s.

When Britannia Airways suffered reduced profits in 1990, the airline made the decision to retire some of the older maintenance intensive 737-200s and replace them with the more efficient 757s. In 1991 four 757s were taken on short-term leases, pending delivery of the new aircraft, six being ordered for delivery in 1992. A further six were ordered in August 1991, the 757s eventually replacing all the 737s. The first 757-200, G-BYAC, was delivered to Britannia on 10 April 1992, with two more arriving in May ready for the busy tourist season.

Iberia was a regular supporter of Airbus since the early days, but Boeing was successful in achieving an order worth $1 billion for 16 757s in June 1990, with options on a further 12 aircraft. The cabin was configured with 26 business class and 102 tourist class seats, and the first delivery (EC-FTR) came on 7 June 1993. At least 12 757s are now in service with further deliveries continuing.

Continental Airlines ordered its first 757s in October 1990 with contracts for 25 aircraft, plus 25 options. The 757s were for use on the carrier's high-density medium- to long-haul domestic services, with expansion on to some international routes. Continental now has a fleet of 33 757s with five more to be delivered during 2000.

F-22 FLYING TESTBED

In mid 1990 the Boeing-owned Number One 757, N757A was modified as a flying laboratory to test the avionics fit for the Lockheed Martin/Boeing/GD YF-22A advanced tactical fighter for the USAF. Some 32 missions were flown over a period of four months to help develop the systems before the maiden flight of the fighter prototype. Further modifications were made in June 1997 when a representative F-22 Raptor nose section, 2.7m long, was fitted forward of the cockpit bulkhead housing the APG-77 radar. Flight testing is continuing with a representative 9m span sensor wing mounted on the 757 fuselage above the cockpit. The fuselage was fitted out with a 30-seat laboratory containing the full range of electronic warfare, communications, navigation and identification sensors to allow full avionics testing to start in August 1998, while the F-22 development aircraft concentrated on more relevant tasks.

The first production 757, originally delivered to Eastern in August 1983, was acquired by NASA in December 1994 and re-registered N557NA. It was used from August 1995 for an initial programme of hybrid laminar flow control (HLFC) system flight trials under contract to NASA. The trials indicated the possibility of cruise fuel consumption savings of between 5 and 15 percent. The test piece was located on a 6.7m section of the wing leading edge, where boundary layer was sucked through 19 million laser drilled holes of up to 0.0125mm diameter to reduce turbulence flow. The savings in fuel costs had to be balanced against the additional costs of manufacture and maintenance of the HLFC system. N557NA was then allocated to flight testing of the YF-23 Advanced Tactical Fighter avionics, competing with the YF-22A, (the project eventually selected by the USAF). This 757 then contributed to the development of the fly-by-wire control system for the Boeing 777.

The Gulf crisis and the resulting worsening world economic situation made 1991 a bad year for air transport. There were marked reductions in air travellers worldwide and, as a result, a

ABOVE: Newcastle-based charter airline Inter European leased two new 757s, the first being G-IEAB (c/n 24636, l/n 259) from AWAS. It served from 1 February 1990 until the airline ceased operations in October 1993 and then joined Airtours as G-LCRC. The second aircraft was delivered new in April 1992 and a third was leased secondhand in April 1992. *Boeing*

number of orders were deferred or cancelled. Many airlines were having difficulties and the first major casualty was Eastern Airlines, which had been experiencing management and union problems for some years. After operating under the protection of Chapter 11, operations finally ceased on 18 January 1991 after 64 years service. Facilities, aircraft and Miami, Atlanta and other airports were abandoned overnight and in many cases just left. While some aircraft were ferried to storage in the desert, others were still on the abandoned gates at Atlanta eight months later with no operators prepared to acquire the assets. The 757s eventually found new operators mainly in the USA and Britain.

Air Europe was forced to suspend operations on 8 March 1991 with orders still outstanding for 12 757s on lease. Although the aircraft allocated to the UK part of the airline were returned to the lessors, and re-allocated to other operators, the independent Italian and Spanish divisions were able to continue operations with 757s in the holiday charter business.

Ethiopian took delivery of the first of five 757s (ET-AJX) on 25 February 1991 — four being in the passenger configuration, and the other one a Package Freighter. US Air, now US Airways, ordered 15 RB.211-powered 757s in November 1991 with options on a further 15. To start operations quickly, the

BELOW: Boeing 757-208ER C-GNXI (c/n 24367, l/n 208) was originally delivered to Odyssey as C-GAWB on 1 February 1989 until operations ceased in June 1990. It was then leased to Nationair on 2 June 1990 until they also ceased flying in April 1993. The aircraft then operated with Air Transat, Leisure Air and Transaero before joining Flying Colours as G-FCLG in February 1998. *Bruce Malcolm*

BOEING 757

ABOVE: Icelandair ordered three RB.211-535E4-powered 757s on 19 October 1988 with the first delivery on 4 April 1990 of 757-208 TF-FIH (c/n 24739, l/n 273). The latter is seen here taking off after hand-over. *Boeing*

airline had already leased 10 of the ex-Eastern 757s which were prepared for service entry in early 1992. They were followed by the new aircraft from 1993, the first being N610AU, delivered on 26 February. Also at the end of 1991, Shanghai Airlines added five more PW2000-powered 757s to the three already in operation. By the end of 1991 total sales of the 757 had reached 770 aircraft, of which 403 with 35 airlines were-powered by the RB.211 turbofans.

MORE ORDERS FOR ROLLS-ROYCE

In an unexpected move in early 1992, UPS selected Rolls-Royce RB.211-535E4 turbofans for a second order for 20 757PFs, even though the initial 20 were P&W-powered. In addition UPS reserved 41 options for Rolls-Royce-powered aircraft. The global economic depression caused difficulties for the traditionally strong carriers, Delta AirLines was an example, with an announcement in April 1992 of major fleet reductions, which could effect the 16 757s still to be delivered, or more likely the outstanding 40 options. The Colombian airline, Avianca took delivery of the first 757, EI-CEY on lease from the GPA Group on 13 August 1992 with a second 757 the following month.

In July 1992 the P&W PW2000-powered 757 was granted 180 minutes ETOPS approval from the FAA, and it was celebrated by American Trans Air with a six-hour flight from Tucson, Arizona to Honolulu in Hawaii. With the opening up of China, Boeing had made significant sales, and on 12 August 1992 the 100th Boeing jet airliner — 757 B-2819 for Xiamen Airlines —

BELOW: Zambia Airways leased 757-23APF 9J-AFO (c/n 24635, l/n 258) from AWAS for routes to Europe, Africa and the Middle East. It was delivered on 18 October 1990 and remained with the airline until October 1993. Following lease to Gulf Air, the aircraft joined DHL in April 1996 based at Bahrain. *Boeing*

IN SERVICE

ABOVE: Ladeco of Chile took delivery of the first of two PW2037-powered 757s on lease from ILFC, CC-CYG (c/n 25044, l/n 369) on 28 May 1991. The aircraft were used on international services from Santiago. The first 757 remained in service until May 1996, when it was leased to Far East Air Transport as B-27001, and the second 757 left the airline in March 1997. *Boeing*

was delivered, with three more to the emerging Chinese regional airlines in the same month. By mid-year Boeing still had an order backlog for 56 jet airliners for China, of which about half were 757s. However, with sales during 1992 of only 15 757s, Boeing was considering reducing the production rate from 8.5 per month, to seven by mid-1993, and to five per month by the end of the year. In December 1992, United announced cancellations of a number of Boeing airliners, but the 757s were unaffected. United already had 71 757s in service, out of a total of 90 ordered, and it was expected that a further four options would be converted into firm orders. ILFC placed a major order for Boeing airliners, including 19 757s with options on five, as well as additional orders for 767s and 777s.

In January 1993, Britannia ordered seven more 757-200ERs to replace the last of the 737-200s, with deliveries starting in May 1994. This order added to the seven 757s already in production for Britannia, with the first aircraft, G-BYAF delivered on 13 January 1993 and G-BYAG before the end of the month. The Dutch charter operator, Transavia took delivery of their first 757, PH-TKA on 22 February 1993, with the second one in May, the fleet eventually growing to four aircraft. On 10 March Air Seychelles doubled its Boeing fleet when Rolls-Royce-powered 757-200ER S7-AAX was delivered for services to Johannesburg and Singapore.

MAJOR CONTINENTAL ORDER

In May Continental Airlines placed a significant order for 92 Boeing jet airliners with 98 options, including 50 757s along with 767s and 777s. In the same month, Ambassador Airways started operations with two second-hand 757s from Newcastle to Cyprus and the Greek holiday resorts. In November UPS ordered 10 more RB.211-powered 757PFs to add to the 55

BELOW: Sterling Airways of Denmark ordered three 757s on 13 January 1989, the first example being OY-SHA (c/n 25155, l/n 371) – seen here departing on its delivery flight after hand-over on 7 June 1991. When the airline ceased operations, the aircraft was repossessed in September 1993 and is now N757AF with Vulcan Northwest. *Boeing*

BOEING 757

ABOVE: With the economic difficulties caused by the Gulf crisis a number of airlines found their earnings reduced drastically. One of the casualties was Air Europe which suspended operations on 8 March 1991. Air Europe 757-236 G-BLVH (c/n 23227, l/n 57) was delivered on 26 March 1985 and has been with Air Europa as EC-GCB since April 1995. It is seen here on approach to London Gatwick in June 1988. *Philip Birtles*

already ordered or in service. Finally, at the end of the year, ILFC added to its order book with five more 757s, bringing total sales for the 757 for the year to 40 aircraft.

Northwest had found 1993 a difficult year, and in January 1994 announced major cancellations and order deferrals. Although no 757s were cancelled outright, 40 deliveries were delayed for an average of 3.5 years, and options on 40 757s were cancelled. However, in early 1995, Northwest decided to bring forward the deliveries of 15 757s to 1995 and 1996, but the balance of 25 were deferred until 2003 to 2005. The airline already had 33 757s in service on US domestic routes.

With the dramatic change of regime in the former USSR, the national airline Aeroflot, which had been without domestic competition, suddenly found a number of new private airlines had formed. Although the initial equipment was often of Russian manufacture, Transaero was the most successful of these start-up carriers, and it took delivery of the first of two leased 757s (EI-CJX) on 1 April 1994, with the second aircraft arriving two weeks later. Two more leased 757s were delivered in May and June 1995 for services to Berlin and Frankfurt. All four 757s were later delivered to Flying Colours and have now

BELOW: TAESA of Mexico has operated a number of 757s one of which was 757-2Y0 XA-TAE (c/n 25268, l/n 400), leased from GPA and delivered on 19 October 1991. It remained in service with the airline until January 1995 when it passed to Air Transat as C-GTSU in November 1996. It was leased to BA as G-CPEP from April 1997. *Boeing*

IN SERVICE

ABOVE: Ethiopian ordered four PW2040-powered 757-200ERs and one 757-200PF on 9 June 1989 with the 200PF ET-AJS delivered first on 24 August 1990, followed by the first passenger aircraft ET-AJX on 25 February 1991. *Boeing*

become part of the jmc fleet. In June Baikal Airlines leased a 757 from ILFC for services from Irkutsk to Moscow.

In mid-1994 Air China was reported to be negotiating for up to 15 757s amongst a number of other Boeing airliners, but in July an 18-month moratorium was put on all new airliner orders while the national industry infrastructure was put in order following a disturbingly high number of accidents. During the first half of the year, one 757 had been delivered to each of China Southwest and China Southern, with a further 11 757s still to be delivered to China Southwest. American Trans Air (ATA), who already had six 757s in service, signed letters of intent with ILFC for two 757-200s and for six aircraft with Boeing.

MORE US BUSINESS

With the world economy improving, in April 1995 United topped up its fleet with orders for five 757s and two 747-400s for delivery by mid 1996, adding to the 88 757s already in service. On 15 September 1995, Lineas Aereas Privadas Argentinas (LAPA) took delivery of 757 LV-WMH followed by the lease of a second hand Lan Chile 757 from 31 March 1997. In October 1995 one 757 and two 767-300ERs were ordered by Uzbekistan Airways, and the National Civil Aviation Authority of Turkmenistan ordered two 757s to add to the one already in service.

Trans World Airlines (TWA) has been struggling for a number of years, surviving financial difficulties, and at the same time reducing in size. Some of the excess costs were being incurred with the operation of older maintenance intensive aircraft, and it was with relief that in February 1996 TWA ordered 20 P&W-powered 757s with options on 20 more, to replace 14 TriStars and some 727s. Ten of the 757s were purchased directly from Boeing and the balance leased from ILFC. Deliveries were scheduled for three aircraft in 1996 and 12 in 1997, to replace TriStars on US domestic services. Two were planned for delivery in 1998 and the remaining three in 1999 to start the replacement of 727s. TWA was therefore replacing three-engine, three-crew aircraft, with more efficient twin-engine, two-crew 757s. The first aircraft, N701TW, an ILFC lease example was delivered on 22 July 1996 and entered service on 1 August between St Louis and the noise sensitive Orange County in California.

Flying Colours, a Manchester-based charter airline was launched in April 1996 with an order for four RB.211-535E4-powered 757s on lease ready for operations to start for the 1997 season. The plan was for three of the aircraft to be based at Manchester, and the other one to be operated from London Gatwick. The first 757, G-FCLA, was delivered on 26 February 1997, with the others arriving at monthly intervals. This airline has been very successful in a highly competitive market and in June 1998 was taken over by Sunworld, who operated Airworld. The Flying Colours title and most of the management team were retained. In 1999 Flying Colours was merged with Caledonian Airways to become jmc AIR with a total fleet of 12 757s.

Having originally shown interest in the 757 in March 1981, Finnair finally selected four 757s in April 1996 for charter operations — the first Boeing airliners to be ordered by the airline. The cabins were configured with 219 seats, and the aircraft are flown on direct operations to the Canary Islands, the Middle East and India. Finnair also has plans to expand 757 operations to transatlantic flights. The aircraft were leased from ILFC with the first delivery, OH-LBO, arriving on 7 October 1997. It was followed by the second aircraft later the same month, and this allowed services to start on 24 October. At about the same time, United ordered five more Pratt & Whitney-powered 757s in May 1996 as part of a contract covering a number of Boeing airliners.

CONDOR LAUNCHES 757-300

With the launch order by Condor, the charter subsidiary of Lufthansa, for 12 of the new stretched 757-300s in September 1996, with 12 options, Boeing hoped to create new interest in the 757 programme, which had slowed down. Condor required the higher capacity 757s to replace its earlier 757-200s on the busy European holiday routes, and configured the cabins with 252 seats to give an improved level of comfort at a slightly greater cost per passenger. The first 757-300, D-ABOC was delivered to Condor on 5 May 1999. Orders announced by Boeing at the same time included five 757-200s for ILFC, and three more for BA, which were confirmed in 1997 when G-CPEN was delivered on 28 March. Delta also ordered four additional 757-200s from ILFC in mid September 1996 to begin the replacement of 737s on the low fare Delta Express operations. By the end of September total 757 sales had reached 801 aircraft, with 699 in service, leaving a backlog of 102 to be delivered.

ABOVE: Newcastle-based Ambassador Airways started operations with two second-hand 757s in May 1993, but ceased operations in November 1994. Boeing 757-236 G-BUDX (c/n 25592, l/n 453) was originally delivered to Caledonian on 13 May 1992. It passed to Ambassador on 1 May 1993 until repossessed on 30 November 1994. It then became SE-DSK with Sunways from May 1995 until they ceased operations in September 1997. It has been with National Airlines as N521NA since July 1999. *Philip Birtles*

In March 1997 Continental increased their firm orders for RB.211-powered 757s by 16 aircraft, bringing the total fleet to 41 aircraft. American Airlines restructured its orders for Boeing jet airliners in May 1997 with 12 757s being delivered at the rate of one per month starting in June 1998. Purchase rights were maintained on a further 38 aircraft. In June 1997, Icelandair added two more RB.211-powered 757-200s to its fleet, and became the second airline to adopt the stretched 757-300 with an order for two aircraft. Options were taken out on a further eight aircraft with deliveries commencing with TF-FIN on 20 January 1998.

THE 757 IN UNIFORM

Boeing had hoped to sell a military transport version of the 767 to the USAF, as a potential phased replacement for the C-135, based on the civil 707 airframe. However, the USAF selected four 757s to replace the VIP VC-137s with the 89th Airlift Wing at Andrews Air Force Base in Maryland near Washington. With the designation C-32, the first example, C-32B 98-0001 made its maiden flight from Renton on 11 February 1998 and was delivered on 1 June, after fitting out at Boeing Field. Unlike previous military transports, the C-32s were treated as civil aircraft and were configured with 45 seats.

In November 1997, Taiwan-based Far Eastern Air Transport (FEAT) signed a letter of intent for five 757s for

IN SERVICE

ABOVE: Baikal Airlines leased a single 757-2Q8 N321L (c/n 26269, l/n 612) from ILFC which was delivered from Boeing Field on 28 April 1994. It was used to link Irkutsk with Moscow until retured to the lessor in April 1996. The plane has been operated by Avianca since August 1996. *Boeing*

deliveries to start in 1999, plus options on a further five aircraft. The order was confirmed as eight aircraft with the first (B-27011) delivered on 8 December 1998, followed by the second aircraft six days later. Although it did not add to the overall sales total, a Rolls-Royce-powered ex-Airtours 757 was delivered to Greenlandair at its Nuuk base, linking the remote Arctic Circle region with Copenhagen in competition with SAS. The aircraft, TF-GRL, was delivered on 30 April 1998 with operations starting on 12 May — it was later re-registered OY-GRL. Technical support comes from Icelandair, and in addition to scheduled flight from Narsarsuaq and Kangerlussuaq/Sondre Stromford to Copenhagen where it is based, the 757 is also used for Novair holiday charter flights to the Mediterranean holiday destinations at weekends.

In October 1998 TWA increased its commitment to the 757 with an order for four aircraft from Boeing, and one from ILFC. The five additional aircraft were scheduled to be delivered by the end of 1999, replacing the last of the airline's 727s.

BELOW: Flying Colours, a Manchester-based charter airline took delivery of the first of four 757s of its initial order on 26 February 1997. The airline absorbed Caledonian Airways for the 2000 season at the same time as the company renamed itself jmc AIR. Boeing 757-25F G-FCLD (c/n 28718, l/n 752) was delivered on 25 April 1997 and is seen here landing at Manchester in October 1999 shortly before repainting in the new jmc colours. *Philip Birtles*

BOEING 757

ABOVE: Iberia ordered 16 757s in June 1990 with options on a further dozen aircraft. The first delivery was EC-FTR on 7 June 1993, and a more recent arrival was 757-256 EC-HDU (c/n 26253, l/n 902) which was delivered on 7 December 1999 and is seen on approach to London Heathrow in February 2000. *Philip Birtles*

In May 1999, British Airways announced plans to reduce its fleet of 53 757s by 10 to 18 aircraft due to falling yields on the short-haul routes. Although the original intention had been for the 757 to be the smallest BA aircraft on the services into London Heathrow, this was revised to transfer some 737-400s from European Operations Gatwick (EOG) as the new Airbus A319s and A320s were delivered to from 2001. With a seating capacity of 200 passengers in the 757s, the substitution with 737s would reduce capacity by some 30 percent, and provide the opportunity to launch an all-business class service on some routes. In additional to the falling yields, increasing capacity problems at Heathrow are a major factor, particularly with continuing delays in the building of Terminal 5.

NEW FREIGHTER CONVERSIONS

By July, BA was in discussions with express package carrier DHL for the sale of about half its 757 fleet for conversion by Boeing into 757 freighters, launching a new modification programme. The total value of the deal, including conversion, was around $500 million. Over half the BA 757 fleet were at least 10 years old, with the oldest approaching 18 years. Many of the earlier aircraft are still-powered by the original lower thrust RB.211-535C turbofans, which would probably be replaced during the conversion. The freighter conversions will be the responsibility of Boeing's Wichita Division with each conversion package costing some $6.8 million. Once the go-ahead was approved the initial conversion is expected to take 12 to 18 months, including certification. DHL plans to allocate the converted freighters to their European organisation, Brussels-based DHL International/European Air Transport, gradually replacing 30 727 freighters.

On 5 October 1999, DHL confirmed that it would be acquiring 44 BA 757s for conversion to freighters, with the first aircraft being delivered mid 2001 after conversion. All 34 RB.211-535C-powered aircraft were included in the deal, plus 10 of the 19 -535E4-powered versions, leaving just nine of the younger aircraft in service. The conversion programme is due to start in August 2000, with all the conversions delivered by the end of 2003. With Boeing Airplane Services (BAS) division at Wichita co-ordinating the programme, and providing additional capacity, much of the conversion work will be done by Israel Aircraft Industries in Tel Aviv and Singapore Technologies in Mobile, Alabama.

In November 1999 it was reported that the newly-formed Manchester based charter airline, jmc AIR were considering the stretched 757-300 for its long term fleet plans, the aircraft being configured with some 280 seats. Should the airline decide to order, the initial commitment is expected to be for two

IN SERVICE

ABOVE: The Dutch charter operator Transavia ordered its first two RB.211-535E4-powered 757s on 23 May 1989. The first was delivered on 22 February 1993. Boeing 757-2K2 PH-TKC (c/n 26635, l/n 608) was the third aircraft and it was delivered on 12 April 1994. It is seen here at Rhodes Airport in September 1998, wearing the new livery and ready for departure. *Philip Birtles*

aircraft, with deliveries beginning in time for the 2001 holiday season in Europe. Also in early 2000, TWA was reported to be in discussions with Boeing for the possible acquisition of 757-300s with 240 seats to replace the older 767-200s. The 757-300s were being considered for US domestic services, as well as operations in the Caribbean and Mexico. By the end of February TWA had 26 PW2037-powered 757-200s in service. Arkia became the second operator to introduce the 757-300 with entry into service in February 2000, allowing the airline to expand its network and offer more international charter flights.

In March 2000 the Boeing operated 757-330 D-ABOI in Condor colours made a sales demonstration tour of Europe before delivery to the airline. Visits were made to Britannia Airways at Luton and also Manchester where existing 757 charter operators include Air 2000, Airtours and JMC Airlines are based. jmc AIR, born out of the amalgamation of Flying Colours and Caledonian Airways, announced an order for two of the stretched 757-300 for delivery in the spring of 2001, the first in the UK.

On 2 May 2000 an order for 20 757-200s from American Airlines brought the Boeing 757 past the 1,000 sales total to an order book of 1,009 aircraft.

BELOW: The USAF ordered four 757s as C-32s for government VIP air transports, replacing the aged VC-137s at Andrews AFB. Maryland. The first aircraft, Boeing 757-2G4/C-23B 98-0001 (c/n 29025, l/n 783) was delivered on 1 June 1998. *Boeing*

6 Customers

On 6 February 1987 America West placed an initial order for three 757s with three options which were soon confirmed. Boeing 757-225 N907AW (c/n 22691, l/n 155) was the first in a further order for 10 757s, having originally been ordered by, but not delivered to, Eastern. This aircraft was delivered on 10 December 1987 and spent a period on lease to TAESA as XA-TCD. *Boeing*

BOEING 757

AIR 2000

BELOW: Manchester-based Air 2000 was formed in April 1986 when an order was placed for one 757-200 with ILFC and another with Chemco Financial Services, both powered by RB211-535E4 turbofans. The first aircraft G-OOOA, seen departing from Manchester in January 1989, was handed over on 1 April 1987 and arrived at Manchester two days later. The second 757, G-OOOB arrived before the end of the month ready for charter operations to begin on 11 April. The network eventually covered over 15 Mediterranean holiday destinations with an average daily utilisation of 16 hours. Air 2000 works very closely with its Canada 3000 associate, sharing aircraft for peak seasons in Europe and North America. When there is less demand for European holidays in the winter, Canadians are keen to leave behind the bitterly cold climate for the warmth of Florida, California and Mexico. Boeing 757-28A G-OOOA (c/n 23767, l/n 127) first flew on 18 March 1987 and has been used by Canada 3000 as C-FOOA. *Philip Birtles*

ABOVE: Air 2000 has now adopted a distinctive multi-coloured livery. Boeing 757-2Y0 G-OOOU (c/n 25240, l/n 388) is leased from GPA, and was delivered on 30 August 1991. It is seen here on finals to Rhodes in August 1997. The Air 2000 fleet has now grown to 13 757-200s, with five more operated by Canada 3000. In addition to flying from Manchester, Air 2000 also has had a main operating base at Gatwick since 1990. The airline also flies from Glasgow, where services to Florida started in September 1989. *Philip Birtles*

BOTTOM: Amongst the holiday destinations regularly served by Air 2000 are the Greek Islands. Boeing 757-23A G-OOOG (c/n 24292 l/n 219) was delivered to the airline on 29 March 1989 and is seen here at Rhodes Airport preparing for departure in September 1995. If the wind is in the right direction, the Airport View Taverna is an ideal place to be for photography. This 757 has also operated with Canada 3000 as C-FOOG. *Philip Birtles*

CUSTOMERS

AIRTOURS

ABOVE: Airtours is another British Manchester-based holiday charter airline which began operations with a number of MD-80s in March 1991. The airline has now standardised on the 757-200. 757-23A G-LCRC (c/n 24636, l/n 259) was leased from Ansett Worldwide Aviation Services and was originally delivered to Inter European Airways as G-IEAB on 1 February 1990. The aircraft is seen here at Manchester in February 2000 being prepared for engine runs following maintenance. *Philip Birtles*

LEFT: In addition to its main base at Manchester, Airtours also has a major operating base at Gatwick, serving European holiday destinations from both airports — as well as a number of other UK airports. Airtours 757-225 G-PIDS (c/n 22195) was the sixth aircraft built and was originally delivered to Eastern on 20 May 1983 as N505EA. It served with Eastern until January 1991. It was delivered to Airtours on 9 January 1995 (after storage in the Nevada desert) and is seen here at London Gatwick, in September 1998. The Airtours fleet of 757s has now reached six aircraft, one of which was delivered new from Boeing in September 1997. The remainder have been acquired secondhand. *Philip Birtles*

AIR TRANSAT

Air Transat is a Montreal Canada-based charter operator which started transatlantic charter flights from Canada to Europe in 1989. The company began replacing its Lockheed TriStars with Boeing 757s in the early 1990s. Boeing 757-23A C-GTSE (c/n 25488, l/n 471) was leased new from AWAS and delivered on 20 November 1992. The airline now has five 757s in operation, and C-GTSE is seen ready for departure from London Gatwick in July 1995. *Philip Birtles*

BOEING 757

AMERICAN AIRLINES

BELOW: American Airlines placed its first order for 50 RB211-535E4 powered 757-200s, with 50 options, in May 1988. The Dallas-Fort Worth based airline now operates a fleet of 102 757s, mainly on domestic North American services. The first aircraft, N610AA, was delivered on 31 July 1989. Boeing 757-223 N661AA (c/n 25295, l/n 423) arrived on 30 January 1992. *Gary Tahir*

ABOVE: This newly-delivered American Airlines 757-223 N641AA (c/n 34599, l/n 351) was handed over on 25 March 1991, and is seen here moving to the departure point at San Diego in April 1991. The last of its current batch of 102 757-200s was delivered to American in May 1999. In May 2000 American signed for another 20 757s, taking the airliner past the 1,000 orders milestone. *Nick Granger*

CUSTOMERS

AMERICA WEST

ABOVE: A number of airlines use their aircraft as flying bill-boards, or to commemorate special events, and America West leads this trend. RB211-535-E4 powered 757-2S7 N905AW (c/n 23567, l/n 97) was originally delivered to Republic Airlines on 19 May 1986, but became surplus when that airline was taken over by Northwest, which operated its own fleet of PW2037-powered 757s. The aircraft was acquired by American West on 11 May 1987 and is carrying Ohio state colours. *Gary Tahir*

RIGHT: In early 1987, Phoenix-based America West placed its initial order for three RB211-535E4 powered 757-200s, with options on a further three. These were followed in December 1988 by 10 more 757s and 15 options. The 757s are used on domestic operations and the first 757-200s were delivered in mid 1987. The airline survived some difficult financial problems, and in the end took delivery of just 15 757s, with 13 currently in service — some of which were acquired second hand. Boeing 757-2G7 N908AW (c/n 24233, l/n 244) was delivered new on 25 August 1989 and is seen ready here preparing for departure from Seattle/Tacoma (Seatac) Airport in May 1998. *Philip Birtles*

AMERICAN TRANS AIR

RIGHT: American Trans Air was formed as a holiday club in August 1973 and became a US national carrier in 1984. It is now a well-established domestic and transatlantic charter airline, based at Indianapolis with a fleet of nine 757-200s — and a further four on order. In May 2000 ATA announced that it would become the North American launch customer for the 757-300, with an order for 10 aircraft. Boeing 757-212 N752AT (c/n 23128, l/n 48) was originally delivered to Singapore Airlines as 9V-SGN on 12 December 1984. Acquired by ILFC, it was delivered to ATA on 12 June 1990 until returned in October 1996, when it was bought by Delta. It is seen here landing at London Gatwick. *Bruce Malcolm*

BOEING 757

CUSTOMERS

ARKIA
LEFT, INSET: The Israeli charter airline Arkia has leased 757-200s in the past from El Al to cover busy holiday periods. Boeing 757-258 4X-EBM (c/n 23918, l/n 156) was delivered to El Al on 17 December 1987, and was leased to Arkia from April to October 1993 with the El Al titles and logos painted out. This aircraft is seen during September 1993 at Rhodes Airport, a popular Israeli tourist destination. Arkia operates one 757-200, and is in the process of taking delivery of two of the stretched 757-300s. *Philip Birtles*

AVIANCA
MAIN PICTURE: Produced to the BA specification, Boeing 757-236 G-BUDZ (c/n 25593, l/n 466) was delivered to Caledonian Airways on 25 June 1992. It was operated by Ambassador Airways from May 1993, until the airline ceased operations in November 1994. During its time with Ambassador, the aircraft was sub-leased to Avianca in Columbia from November 1993 until April 1994. It is seen here at London Gatwick in January 1994 on a charter. Following the closure of Ambassador, this 757-200 was registered SE-DSL with Sunways Airlines, a Swedish holiday charter airline who also ceased operations in September 1997. The aircraft was leased to Royal Airlines as C-GRYK in March 1998. *Nick Granger*

BELOW, INSET: Avianca operates a fleet of four 757-200s to mainly US destinations. Boeing 757-2Y0 EI-CEZ (c/n 26154, l/n 486), leased from GPA, was delivered on 22 September 1992. It is seen here at Miami in January 1994. *Nick Granger*

BOEING 757

ABOVE: Britannia Airways has a major operating base at London Gatwick, and conducts operations from a number of other UK airports. Its 757-200 fleet serves all the major European holiday destinations. Boeing 757-204 G-BYAC (c/n 26962, l/n 440) was the first new 757 to be delivered on 10 April 1992, arriving at Luton the following day. The aircraft is seen here on arrival at London Gatwick on 20 April 1992. This aircraft was leased to Istanbul Airlines as TC-ARA, and is currently with National Airlines as N512NA. *Philip Birtles*

BRITANNIA AIRWAYS

ABOVE: Britannia Airways has maintained its headquarters and engineering base at London's Luton Airport since the airline was formed in December 1961. This major British charter airline placed its initial order for six Boeing 757-200ERs in 1990 to start replacing an older fleet of 737-200s. With those deliveries not due until 1992, the airline took four aircraft from Air Holland on short-term lease for the 1991 season. Boeing 757-23A G-OAHK (c/n 24291, l/n 215) was leased by Air Holland as PH-AHK, from AWAS, and delivered on 2 March 1989. It was leased on to Britannia Airways from November 1990 until June 1998 (with some breaks) until finally returned to Air Holland. The aircraft is seen at Luton in April 1992.
Philip Birtles

BELOW: Blue Scandinavia was a Swedish charter operator, originally equipped with a TriStar and which started operations to the Greek resorts in 1997. Boeing 757-236 SE-DUO (c/n 24795 l/n 279) was delivered to Air Europe as G-BRJI in April 1990. After service with Air Europa, Ambassador Airways and Venus Airlines, it joined Blue Scandinavia on 8 May 1997. That airline was later taken over to become the Scandinavian division of Britannia Airways, in April 1998. The 757 is seen here in the Blue Scandinavia colours, ready for departure from Rhodes Airport, in August 1997. *Philip Birtles*

CUSTOMERS

BRITISH AIRWAYS

ABOVE: British Airways shared the honour of launching the 757 with Eastern Airlines. Its aircraft have been used largely on the UK Shuttle routes from Heathrow, as well as lower density services to European destinations. Many of the earlier BA 757s are still powered by the RB211-535C engines, now the only versions of the engine still in operation. Because the aircraft have been in service with BA for so long, they have gone through four basic changes of liveries. Boeing 757-236 G-BIKI (c/n 22180, l/n 25) was delivered on 30 November 1983 and named *Tintagel Castle*. It is seen here on approach to London Heathrow in the initial 'British' livery with the Speedbird above the cockpit. *Philip Birtles*

BELOW: British Airways next adopted a fairly conservative colour scheme with a shield and coat of arms on the fin, with the top of the aircraft painted light grey. A number of the BA fleet are still painted in this scheme while awaiting their turn for a repaint in the latest 'ethnic' colours. BA 757-236 G-BIKS *Corfe Castle* (c/n 22190, l/n 63) is seen here at London Heathrow, in March 1988 ready for departure. *Philip Birtles*

BOTTOM: In mid-1997 British Airways launched the controversial 'Citizens of the World' livery, using some 50 international artists to create fin designs to project a global image. One of these is seen on 757-236ER G-BPED (c/n 25059, l/n 363), seen here on approach to London Heathrow in August 1998. *Philip Birtles*

BOEING 757

CANADA 3000
ABOVE: In co-operation with Air 2000, Canada 3000 operates a fleet of five 757-200s for holiday charter flights to the southern states of the USA and the Caribbean, particularly during the cold Canadian winter. Canada 3000 retains a livery similar to that used by its UK partner. Boeing 757-28A C-FXOF (c/n 24544, l/n 280) is leased from ILFC and was delivered on 30 April 1990. *Canada 3000*

CHALLENGE AIR CARGO
MAIN PICTURE: Challenge Air Cargo operates a fleet of three Boeing 757-200PFs from its Miami base. With the plain white-painted fuselage, it is possible to see the 757's slight double-bubble fuselage cross-section, with the join line at cabin floor level. Boeing 757-23APF N573CA (c/n 24971, l/n 340) was originally delivered to Gatwick based Anglo Air Cargo as G-OBOZ on 19 August 1991. However, that airline ceased operations in January 1992 and the aircraft was delivered to Challenge Air Cargo on 2 February 1992. It is seen on approach to Miami in January 1994. *Nick Granger*

CHINA SOUTHERN AIRLINES
ABOVE RIGHT: As China gradually emerges into a more open economic environment, air travel has increased substantially on domestic and international routes. In October 1987 the single-entity government airline CAAC ordered three RB211-535E4 powered 757-200s, with three more following a year later. With the reorganisation of commercial aviation in China, CAAC became responsible purely for the administration of civil aviation and a number of new airlines were formed around the country. Amongst China's international, regional and domestic operators is Chengdu-based China Southwest Airlines which operates 12 757s. One of the largest is Shanghai-based China Southern which operates international, regional and domestic services with a fleet that includes 18 757-200s. The first of these was delivered in September 1987. 757-21B B-2806 (c/n 24401, l/n 232) was delivered to China Southern on 28 August 1989, and is seen here on finals to Hong Kong Kai Tak in April 1998. *Philip Birtles*

CUSTOMERS

BOEING 757

CONDOR
ABOVE: Condor is the charter subsidiary of Lufthansa, operating holiday flights to Mediterranean resorts with Boeing 757s and 767s. The airline placed its first order for PW2040-powered 757s on 29 September 1988. The first aircraft, D-ABNA, was delivered on 19 March 1990 and today the fleet includes 18 757-200s — some of which are now being replaced by the new 757-300s. Boeing 757-230 D-ABNR (c/n 26434, l/n 532) was delivered on 16 March 1993 and is seen here at Rhodes Airport in September 1995. *Philip Birtles*

BELOW: At the Farnborough Air Show in September 1996 Condor placed the launch order for the stretched 757-300 with a contract for 12 aircraft, and options on a further 12. The first 757-300, D-ABOE was delivered on 10 March 1999. D-ABOG (c/n 29014, l/n 849), which was was delivered on 19 March 1999, is seen here at Gatwick in February 2000 undergoing work on its RB211-535E4 turbofans. *Nick Granger*

CONTINENTAL AIRLINES
ABOVE, RIGHT: Texas-based Continental Airlines placed its initial order for 25 757-200s on 8 October 1990 to serve its US domestic routes. The fleet has now grown to 33 aircraft, with four more to be delivered. The first 757-200, N58101 was delivered on 12 May 1994. Continental 757-224 N12116 (c/n 27558, l/n 702) was delivered on 27 March 1996 and is seen here on a visit to London Gatwick in July 1999.
Nick Granger

CUSTOMERS

DELTA AIRLINES
MAIN PICTURE: Atlanta-based Delta AirLines uses a fleet of 102 757-200s on US domestic routes, with 17 still to be delivered. Delta first ordered an initial 60 757-200s on 12 November 1980, launching the PW2037 turbofan on the 757. Delta is in the process of introducing a new livery to the fleet. Boeing 757-232 N604DL (c/n 22811, l/n 43) was delivered on 3 December 1984 and is seen here in the new livery shortly after take-off. *Delta*

BOEING 757

DHL
ABOVE: DHL Worldwide Express has operated a dedicated fleet of 757-200 Package Freighters for some years. DHL has now signed a contract with Boeing and British Airways to take delivery of the majority of the BA 757 fleet and convert them to freighter configuration. Seen at Luton in April 2000 DHL 757-23APF VH-BRN (c/n 24868, l/n 314) was originally delivered to Challenge Air Cargo on 24 September 1990. *Philip Birtles*

EL AL
BELOW: El Al, the Israeli flag carrier ordered five Boeing 757-200s on 1 October 1986, to operate on its regional routes. The first 757-200 4X-EBL was delivered on 25 November 1987 and the fleet now includes eight 757s, which are regularly used on the less busy European services. Boeing 757-258 4X-EBR (c/n 24254, l/n 185) was delivered on 19 July 1988 and is seen here arriving at Manchester in January 1989. *Philip Birtles*

In addition to scheduled operations, El Al also uses its 757s for holiday charter flights around the Mediterranean resorts — the Greek island of Rhodes being particularly popular. Boeing 757-258 4X-EBM (c/n 23918, l/n 156) was delivered on 17 December 1987 and is seen here on arrival at Rhodes Airport in September 1998. *Philip Birtles*

CUSTOMERS

ETHIOPIAN AIRLINES
ABOVE: Ethiopian ordered four PW2040 powered Boeing 757-200s and one Package Freighter on 9 June 1989, with the first delivery, ET-AJX on 25 February 1991. The latter aircraft is seen here on the flight line at Renton being prepared for hand over to its new owner. *Boeing*

FINNAIR
Finnair first showed interest in the 757 in March 1981, but did not place an order until April 1996 — with a contract for four aircraft. These were the first Boeings to be bought by Finnair. The first 757, OH-LBO (c/n 28172, l/n 772), is seen here on the Renton flight line being prepared for delivery on 7 October 1997. *Boeing*

BOEING 757

Finnair 757-2Q8 OH-LBS (c/n 27623, l/n 792) was delivered on 9 March 1998 and is seen at Arrecife, in the Canary Islands, in January 1999 — carrying Father Christmas and his reindeers by way of Christmas decoration. *Nick Granger*

GREENLANDAIR

ABOVE: Greenlandair operates a single Boeing 757-200 on scheduled services between Greenland–Copenhagen, with support from Icelandair. On some weekends the same aircraft operates Novair charters to Mediterranean holiday destinations. Boeing 757-236 TF-GRL (c/n 25620, l/n 449) was originally delivered to Britain's Inter European Airlines as G-IEAC on 28 April 1992. When that airline ceased operations, the 757 was acquired by Airtours as G-CSVS in October 1993, and finally moved to Greenlandair in April 1998 as TF-GRL. It was re-registered as OY-GRL in April 1998 and is seen here ready for departure from Rhodes Airport in September 1998 while still registered TF-GRL. *Philip Birtles*

CUSTOMERS

IBERIA
ABOVE: Iberia placed an order for 16 757-200s with options on 12 more in June 1990. The aircraft are used on a number of European routes and are configured with 86 business- and 102 tourist-class seats. Boeing 757-256 EC-FYL (c/n 26244, l/n 616) was delivered in August 1994, but passed to National Airlines as N508NA in August 1999. *Philip Birtles*

ICELANDAIR
RIGHT: Icelandair placed its initial order for three RB211-535-E4 powered 757-200s on 19 October 1988 for routes to the USA. 757-208 TF-FIH (c/n 24739, l/n 273) was the first to be delivered on 4 April 1990 and is seen here ready for departure from London Heathrow in April 1997. *Philip Birtles*

BELOW: Icelandair unveiled a new corporate image in November 1999, and a new livery appeared on its 757. The airline is planning to expand the current fleet of six 757-200, plus a 757 freighter, with two more 757-200s and two stretched 757-300s ordered in June 1997. Boeing 757-208 TF-FII (c/n 24760, l/n 281) was delivered on 3 May 1990 and is seen in the new colours on approach to London Heathrow in January 2000. *Philip Birtles*

jmc AIR

ABOVE: Caledonian was originally the Gatwick-based charter division of British Airways, replacing British Airtours when British Caledonian was taken over by BA. The airline was sold to Inspirations at the end of 1994 with a fleet of five TriStars and Boeing 757-200s to the same specification to the BA 757-200s (but powered by RB211-535E4 turbofans). The first 757-236, G-BPEA was delivered on 31 March 1989 and the aircraft could be interchanged with the BA fleet. Boeing 757-236 G-BPEB (c/n 24371, l/n 225) was delivered on 27 April 1989 and passed to BA in October 1995. It is seen here ready for departure at Rhodes Airport in September 1995. *Philip Birtles*

ABOVE RIGHT: Manchester-based Flying Colours was launched in April 1996 with an order for four new RB211-535E4 powered 757-200s for charter flights starting in the 1997 holiday season. Three aircraft were based at Manchester and the others at Gatwick. The airline merged with Airworld in June 1998, but retained the Flying Colours image and most of the senior management team. Boeing 757-25F G-FCLD (c/n 28718, l/n 752) was delivered on 25 April 1997 and is seen climbing away from Manchester in October 1999 soon to be painted in the new jmc livery. *Philip Birtles*

MAIN PICTURE: On 1 September 1999 Flying Colours was re-branded as jmc AIR when Caledonian and Flying Colours were merged. The Caledonian 757s were returned to BA and the 12 757-200s of Flying Colours were repainted in the new colours. The main jmc AIR base is Manchester, with another at Gatwick, and further services from a number of other UK airports. One of the first 757-200s to be repainted was G-FCLD seen here at London Gatwick just before Christmas 1999 ready for the 2000 holiday season. *Nick Granger*

CUSTOMERS

LTU

ABOVE: Dusseldorf-based LTU ordered nine RB211-535E4 powered 757-200s on 25 August 1983 with the first delivery, D-AMUR, on 25 May 1984. The fleet has now increased to 12 757-200s together with other types. The aircraft are operated on non-scheduled holiday flights throughout Europe, particularly to the Mediterranean resorts. Boeing 757-2G5 D-AMUQ (c/n 26278, l/n 671) was delivered on 26 April 1995 and is seen here on finals to Rhodes in September of the same year. *Philip Birtles*

LEFT: The German subsidiary LTU SUD shares the operation of the 757-200s with a revised livery. Boeing 757-2G5 D-AMUM (c/n 24451, l/n 227) was delivered to LTU Sud on 5 May 1989 and is seen here from the Airport View Taverna ready for departure from Rhodes in September 1995. *Philip Birtles*

BELOW: The Spanish division of LTU is the Majorca-based LTE with a fleet of three 757-200s. LTE 757-225 EC-ETZ (c/n 22689, l/n 117) was originally delivered to Eastern as N525EA on 19 December 1986. It was acquired by LTU for LTE in February 1990, returning to LTU as D-AMUK in October 1993. *Philip Birtles*

CUSTOMERS

MEXICANA

RIGHT: Mexicana introduced the first of two PW2037-powered 757-200s on lease from ILFC in January 1997. This total was later increased to five aircraft, for services from Mexico to North American destinations. The first aircraft was 757-2Q8 N55MX (c/n 24964, l/n 424) originally delivered to American Trans Air on 5 February 1992, and it served there until November 1996. It was delivered to Mexicana on 3 December 1996 ready for operations to commence a month later. *Gary Tahir*

MONARCH AIRLINES

ABOVE: Monarch was the first charter airline to opt for the 757-200 when the Luton-based operator placed an initial order for two aircraft on 19 February 1981 powered by RB211-535C turbofans (later replaced by the more economical -535E4s). Monarch had started life at Luton in 1968, with Bristol Britannias for passenger and cargo charters. The Boeing 757s replacing elderly Boeing 720s. Monarch's sixth 757-2T7 G-DRJC (c/n 23895, l/n 132) was delivered on 12 May 1987, entering service two days later. This was the second aircraft to carry the initials of a (retired) director in its registration, and it is seen on turnaround at Luton in the first month of service. *Philip Birtles*

BELOW: Although Monarch has its headquarters and main engineering facility at Luton Airport, many flights are operated from Gatwick and Manchester. The latter airport also houses a major additional engineering facility. Boeing 757-2T7ER G-DAJB (c/n 23770, l/n 125) was delivered on 16 March 1987 and is seen here just lifting off from Luton in March 1992. *Philip Birtles*

CUSTOMERS

MYANMAR INTERNATIONAL AIRWAYS
LEFT: Royal Brunei Airlines took delivery of 757-2M6 V8-RBA (c/n 23452, l/n 94) on 6 May 1986. It was first leased to Myanmar International Airways from August 1993 for one year and is seen taking off from Hong Kong Kai Tak in Myanmar markings. *Asian Aviation Photography*

NORTH AMERICAN
BELOW: New York–JFK-based North American includes two 757-200ERs in the fleet. With the inaugural flight of N750NA on 20 January 1990, North American partnered with El Al to specialise in flying their transfer passengers to other major destinations in the USA. This is a great saving for El Al, which otherwise would be flying a part-empty 747 to destinations across the USA. Boeing 757-23A N757NA (c/n 24567, l/n 257) was delivered on 1 March 1990. *North American*

ABOVE: The two North American 757-200ERs are powered by RB211-535E4 turbofans and the cabins can be configured for either 169 tourist passengers, or a mix of 24 first class and 109 tourist. The seats are leather and each passenger has their own video screen. Many flights are operated for the US government to destinations as far away as the Indian Ocean and Manila, with *ad hoc* charters filling the gaps. Boeing 757-23A N757NA is seen at Miami in January 1994. *Nick Granger*

BOEING 757

NORTHWEST

ABOVE: Northwest placed an initial order for PW2037 powered 757-200s on 29 November 1983 with the first delivery, N501US, on 28 February 1985. The Minneapolis based airline now operates a fleet of 48 757s, with 25 more on order. Northwest 757-251 N513US (c/n 23201, l/n 83) was delivered on 11 February 1986 and is seen on push-back at Minneapolis in May 1998. *Philip Birtles*

LEFT: The Northwest 757-200s serve domestic routes from Minneapolis and Detroit, for both Northwest and its partner KLM, feeding international flights. Boeing 757-251 N537US (c/n 26484, l/n 697) was delivered on 20 February 1996 and is seen on approach to Seatac Airport between Seattle and Tacoma in May 1998. *Philip Birtles*

PRIVATAIR

Geneva based PrivatAir is a unique operator of a Boeing 757-200 VIP executive jet for special charters. Originally known as Petrolair, its Boeing 757-23A HB-IEE (c/n 24527, l/n 249), leased from AWAS, was delivered on 2 October 1989 as HB-IHU. It was re-registered HB-IEE in November 1989 and is seen here landing at Luton in September 1991. *Philip Birtles*

CUSTOMERS

ABOVE: Since being renamed PrivatAir in April 1997, 757-23A HB-IEE has been repainted in a smart new livery which is shared by the Boeing 737 and Gulfstream aircraft also in the fleet. The aircraft often makes visits to London Heathrow for maintenance with BA, where it is seen ready for departure in September 1998. *Nick Granger*

LEFT: The PrivatAir 757-200 was booked by The Rolling Stones and their entourage on their 'Bridges to Babylon' European tour in 1999. With a total capacity for 60 passengers, the forward cabin is configured with two dining tables seating four, three two-place tables, a three-seat couch and an arm chair. The mid cabin is furnished with 20 fully-reclining sleeper seats, and the rear cabin has 27 economy seats. *PrivatAir*

ROYAL AVIATION

BELOW: Montreal based Royal is a Canadian charter airline currently with three 757-200s. It began operations in January 1992, using three TriStars on transatlantic flights. Boeing 757-236 C-GRYK (c/n 25593, l/n 466) was originally delivered to Caledonian in June 1992 as G-BUDZ. It spent periods on lease to Nationair Canada as C-FNXY, Ambassador, AVIANCA and Sunways as SE-DSL. It was the first 757 to be delivered to Royal on 17 March 1998 as the start of replacing the TriStars. *Gary Tahir*

BOEING 757

CUSTOMERS

ROYAL AIR MAROC
ABOVE LEFT: RAM ordered two PW2037 powered 757-200s on 5 February 1986, with the first delivery, CN-RMT arriving on 15 July 1986. The aircraft are used for European and Middle Eastern scheduled services. The second 757-2B6 CN-RMZ (c/n 23687, l/n 106) was delivered on 7 August 1986 and is seen here on approach to London Heathrow in December 1988. *Philip Birtles*

ROYAL BRUNEI
ABOVE: Royal Brunei ordered three RB211-535E4 powered 757-200s on 30 May 1985 with the first delivery, V8-RBA occurring on 6 May 1986. Boeing 757-2M6 V8-RBB (c/n 23453, l/n 100) was delivered on 13 June 1986 and is seen here during the following week, on display at the IAS '86 air show, Jakarta. The third 757 was used by the Sultan of Brunei and then sold to Kazakhstan. The remaining 757s are presently leased to Myanmar and Qatar. *Philip Birtles*

ROYAL NEPAL AIRLINES
MAIN PICTURE: Royal Nepal ordered a pair of RB211-535E4 powered 757s on 17 February 1986. One was in standard passenger configuration, while the other was the sole Combi 757 to be produced. The Combi, 9N-ACB was delivered on 15 September 1987. This aircraft, 9N-ACB (c/n 23863, l/n 182), is seen climbing out of Renton for its maiden flight on 15 July 1988, carrying the test registration N5573K. The upward opening side cargo door sill can be seen between the forward passenger doors under the 'Nepal' title. *Boeing*

BOEING 757

CUSTOMERS

SHANGHAI AIRLINES
LEFT: Shanghai Airlines placed its initial order for PW2037-powered 757-200s on 14 December 1988. The first delivery, B-2808, came on 4 August 1989. This 757-26D (c/n 24471, l/n 231), is seen on the Renton flight line being prepared for flight testing. The airline's fleet now includes seven 757-200s. *Boeing*

TRANSAVIA
INSET, BELOW: Dutch charter carrier Transavia placed their initial order for two 757-200s on 23 May 1989, later increasing the Schipol based 757 fleet to four aircraft. The aircraft are used for charter flights to European holiday destinations together with some 737s. Boeing 757-2K2 PH-TKA (c/n 26633, l/n 519) was delivered on 22 February 1993 and is seen in the original livery departing Rhodes Airport in September 1995. *Philip Birtles*

MAIN PICTURE: In the last few years Transavia have repainted their fleet in a new livery. Boeing 757-2K2 PH-TKB (c/n 26634, l/n 545) was delivered on 3 May 1993 and is seen in the distinctive new colours at a typical Mediterranean airport. *Transavia*

TURKMENISTAN AIRLINES

ABOVE: Ashkhabad-based Turkmenistan Airlines operates two 757-200s on its European schedules. 757-22K EZ-A011 (c/n 28336, l/n 725), was delivered on 29 August 1996 and is seen here on approach to Heathrow in February 2000. *Philip Birtles*

RIGHT: Turkmenistan's Boeing 757-22K EZ-A012 (c/n 28337, l/n 726) was delivered on 30 August 1996 and has spent a period on lease to Royal Nepal Airlines. *Philip Birtles*

CUSTOMERS

TRANS WORLD AIRLINES
BELOW: St Louis-based TWA operates a fleet of 21 757-200s, with six more on order. TWA placed its initial order for 10 PW2037-powered 757-200s, from both Boeing and ILFC in February 1996. Boeing 757-2Q8 N706TW (c/n 28165, l/n 743) was leased from ILFC and delivered on 18 February 1997. It is seen here at London Gatwick in June 1998. *Nick Granger*

UNITED AIRLINES
ABOVE: United placed its first order for PW2037-powered 757-200s on 26 May 1988, with the first delivery (N501UA) following on 24 August 1989. The 757 fleet has now grown to 98 aircraft. Boeing 757-222 N509UA (c/n 24763, l/n 284) was delivered on 21 May 1990 and is seen here about to take-off from Boeing's Renton Field. *Boeing*

BOEING 757

LEFT: In recent years United has adopted a new corporate image, using sober grey and blue colours. The airline's 757-200s are used on North American domestic routes. Boeing 757-222 N582UA (c/n 26702, l/n 550) was delivered on 19 May 1993, and is seen here on turnaround at Seatac in May 1998. *Philip Birtles*

UNITED PARCEL SERVICE
Louisville-based UPS launched the 757-200PF into production with an initial order for PW2040-powered Package Freighters on 21 December 1985. However, all subsequent orders have specified the RB211-535E4 turbofans. The 757s are used mainly on North American overnight package flights, from the Louisville hub. The aircraft feature an upward-opening cargo door on the port side of the fuselage, a smaller crew entry door further forward and all the windows have been deleted. The first 757-200PF, N402UP was delivered on 17 September 1987 with the second following the next day. PW2040-powered Boeing 757-24APF N409UP (c/n 23731, l/n 186) was delivered on 29 July 1988, and is seen here at Clearwater, Florida, in November 1989. *Philip Birtles*

CUSTOMERS

US AIRWAYS
ABOVE: US Air became US Airways in February 1997 and its fleet of 757s includes 11 ex-Eastern aircraft, plus 23 new 757s. US Airways has its headquarters at Arlington in Virginia. The airline placed a first order for 15 757-200s on 30 March 1992 and now has a fleet of 34 757-200s. US Air 757-2B7 N623AU (c/n 27244, l/n 607) was delivered on 24 March 1994 and is seen here on approach to Washington National Airport in May 1996. *Philip Birtles*

BOEING 757

US AIR FORCE

RIGHT: Four Boeing C-32s have been ordered by the USAF for service with the 89th Airlift Wing at Andrews Air Force Base, Maryland, near Washington DC. The VIP-configured aircraft replaced the ageing VC-137s (Boeing 707s) used for senior US government and head-of-state transport missions. The first C-32 made its maiden flight on 11 February 1998, from Renton. The first two aircraft were designated C-32Bs, with the first (98-0002) delivered on 29 May 1998. This USAF C-32 99-0003 (c/n 29027, l/n 824) was built to commercial standards as a 757-2G4 and was the first of the two C-32As. It was delivered on 20 November 1998 and is seen here on approach to London Heathrow in April 1999. *Nick Granger*

XIAMEN AIRLINES

BELOW: Based in Xiamen, China, this airline operates domestic and regional services using five 757-200s and is one of the nation's important secondary carriers. Boeing 757-25C B-2819 (c/n 25898, l/n 475) became the 100th Boeing jet airliner to be delivered to China when it was handed over on 12 August 1992. It is seen here at Boeing Field ready for the delivery flight. *Boeing*

BELOW RIGHT: Xiamen's operating base is located near Hong Kong, opposite to the island of Taiwan, and the airline operates regular scheduled services to Hong Kong. Xiamen placed its initial order for three RB211-535E4-powered 757-200s on 27 October 1989, followed by two more as part of a CAAC allocation. Boeing 757-25C B-2848 (c/n 27513, l/n 685), was the fourth 757 for the airline and was delivered on 7 August 1995. It is seen at Hong Kong, Kai Tak in April 1998. *Philip Birtles*

CUSTOMERS

7 Accidents and Incidents

With only five airframes lost from 1983 to the end of 1999, the 757 has proved to be a safe aircraft. It has shown itself to be a reliable operator in all climates — whether in the hands of a major airline with over 100 aircraft, or a modest local carrier with one or two 757s.

FIRST LOSS IN CHINA

The first total loss came on 2 October 1990 when a hijacked Xiamen Airways 737 hit 757 B-2812 while landing at Guangzhou, China. The 757 was standing at the at the runway holding point. It was believed that the hijacker detonated a bomb when he realised that the 737 was not landing at Taiwan as demanded. At least 127 people were killed and more than 60 injured, but at least 40 passengers and crew escaped from the wreckage of the 757, the whole centre-section of the cabin being destroyed. A nearby (empty) China Southwest 707 was also destroyed by flying debris.

AMERICAN 965

The period from late 1995 to the latter part of 1996 was a bad one for 757 losses and passenger fatalities. On 20 December 1995, American Airlines 757 N651AA, flying from Miami to Cali in Columbia, crashed into mountains at night killing all but four of the people on board. The aircraft was descending into Cali Airport when it hit a 3,657m (12,000ft) mountain at a height of around 2,743m (9,000ft) near the town of Buga. The impact was some 18.5km (10nm/11.5 miles) east of the normal let-down path even though all the navigation beacons and aids were working.

There was no immediate indication of the accident's cause, the flight data recorder (FDR) and the cockpit voice recorder (CVR) giving no indication of any technical problems, but the ground proximity warning system was heard to alert the crew too late for any action. Following analysis of the CVR readout, it was established that the crew did not carry out the appropriate pre-descent approach briefing or checks before entering the steep sided valley. The aircraft hit the high ground as the crew established it for the approach and it was amazing that there were any survivors.

BIRGENAIR 301

On 6 February 1996 Turkish charter airline Birgenair 757 TC-GEN departed the Dominican Republic at night in light rain *en route* for Berlin and Frankfurt via Gander. When climbing through 2,133m (7,000ft) the crew called approach control to advise that a return was being made to the airport. After the message 'stand by', no further communications were received from the aircraft, and the 757 was found to have fallen into the sea some 20km (12.5 miles) north of Puerto Plata with the loss of all 176 passengers and 13 crew. With the help of the US Navy, the FDR and CVR were recovered from the seabed by the end of the month, together with other items of wreckage. It soon began to emerge that the captain's airspeed indicator may have been at fault, and a factor in the accident. The FDR showed an indicated airspeed (IAS) of 335kt (620km/h) when the stall warning activated the stick shaker at 2,225m (7,300ft). However, the normal stick shaker speed is 130kt IAS. The stick shaker operated for 83 seconds before the aircraft hit the sea, the pilots failing to recover from the stall.

Following analysis of the cockpit voice recorder, it was realised that the crew were aware that the captain's ASI was not working as it should, and elected to use the first officer's. They failed to consult the appropriate check-list or manuals, nor did they compare the main ASIs with the analogue standby unit. It was found that the aircraft had been parked at Puerto Plata since 23 January without the appropriate pitot head covers being used, causing potential blockage by insects or other debris. Soon after take-off, the captain had engaged the autopilot in the climb. The autopilot is slaved to the air-data computer, which receives airspeed information from the captain's ASI. As the aircraft continued to climb, the captain's ASI began to read higher than the correct airspeed, and as a result the autopilot/autothrottle increased pitch up and reduced power to lower the apparent higher than normal airspeed. Due to crew confusion, manual control was not regained, the aircraft stalled and was out of control for one minute and four seconds before hitting the sea.

AERO PERU 603

Aero Peru 757 N52AW was lost in the Pacific Ocean on 2 October 1996 killing all 70 occupants when flying from Lima in Peru to Santiago in Chile. Shortly after take-off at night and in poor visibility, the crew requested a return to Lima with 'mechanical problems'. After continuing radio communications between the crew and air traffic control for some 30 minutes, all contact was lost and the aircraft was found to have crashed into the sea off Passamayo, some 75km (40nm) north of Lima.

The captain reported losing the basic instrument indications for height and speed, and despite the reduction of power, the aircraft still appeared to be accelerating. During the accident investigation, it was found that during maintenance prior to the flight, the static vents had been taped over for protection, but the tape had not been removed prior to flight, resulting in the loss of control and stall into the sea.

ACCIDENTS AND INCIDENTS

BRITANNIA

The most recent accident, fortunately without loss of life, involved Britannia Airways 757 G-BYAG which aquaplaned and skidded off the right side of the runway at Gerona in northern Spain on 15 September 1999. The aircraft was landing, just before midnight local time, in heavy rain following a lengthy period of thunder storms. The aircraft, with 245 passengers and crew on board, was on a holiday charter from Cardiff. It careered across wet ground and down an embankment before coming to a rest in a field with the fuselage broken into three main sections and engines torn off. Two people were seriously injured. Between the storms the wind was reported as light northerly with visibility at 5km (3.1 miles), reducing to 2km (1.2 miles) in rain. The downsloping runway was 2,400m (7,874ft) long and equipped with an instrument landing system (ILS).

Following the accident investigation, it was found that the aircraft had touched down fast nose-wheel first with a high sink rate. After going around from a VOR/DME approach to runway 02, the captain flew an autopilot linked ILS approach to runway 20. The crew had the runway in sight at 150m (500ft) above ground level (AGL) and surface wind was notified as 150° at 6kt (11km/h). At the decision height of 82m (270ft) AGL, the captain elected to land, but the crew suddenly lost

ABOVE: Britannia Airways 757-204 G-BYAG (c/n 26965, l/n 517) was delivered on 22 January 1993 and was damaged beyond repair when landing during thunderstorm activity at Gerona in north-east Spain. Fortunately there were no fatalities when the aircraft came to a halt with the fuselage broken into three sections. The 757 is seen here about to turn on the runway before backtracking, at Luton in April 1999. *Philip Birtles*

outside visual reference and the high sink rate warning sounded. The thunderstorm activity had not been notified to the crew. With a runway centreline touchdown at 141kt, the force of the landing pushed the nose gear up into the electronics bay below the cockpit. As a result it was believed that control was lost due to engine control damage which may have caused power asymmetry. This 757 was a total write-off.

None of the accidents were attributable to the lack of integrity of the aircraft, engines or systems. The Chinese 757 was simply an innocent bystander to an act of aerial piracy. Insufficient care prior to take off caused the Birgenair and Aero Peru accidents, and in the latter aircraft it would have been difficult to see the tape in the dark without the normal red flags attached to the blanks. It appears likely that the Britannia 757 entered a form of microburst on the final approach to land, which makes control practically impossible and has caused much more serious accidents in the past.

8 Production History

Identification of the Boeing 757 is straightforward as, until recently, there has only been one basic version of the aircraft, the Model 757-200 (with the incorporation of standard Boeing customer numbers). Despite an evolving range of powerplant options from Rolls-Royce and Pratt & Whitney the passenger-carrying 757 remained simply designated as the 757-200 — with no distinction made between those versions with different operating weights. Boeing has introduced improved cockpits in later-production aircraft, but again with no suffix to the variant designator. It is only with the cargo-carrying versions of the 757 that some variations in designation began to creep in. The arrival of the stretched 757-300 marked the first major change in the design's 20-year production life.

BOEING 757 VARIANTS:
757-200: Basic passenger version, can be configured for up to 239 seats
757-200PF: Package Freighter version developed to an order from US carrier, United Parcel Service
757-200 Combi: Combined passenger/freight-carrying version, with reconfigurable main cabin and side cargo door. Has also been referred to as the **757-200M** or **757-200CB**
757-200QC: Proposed Quick Change freighter version, with main deck side cargo door
C-32A: VVIP version of 757-200 developed for US Air Force
757-300: Stretched version of 757-300, with 20 percent more seating space

Boeing 757 Production List

L/N	C/N	Model	Operator (original & current)	Identity	Delivered
001	22212	200	Boeing	N757A	
002	22191	225	Eastern/NASA as N557NA	N501EA	18.8.83
003	22192	225	Eastern/US Air as N600AU	N502EA	28.9.83
004	22193	225	Eastern/US Air as N601AU	N503EA	25.5.83
005	22194	225	Eastern/Airtours as G-JALC	N504EA	28.2.83
006	22195	225	Eastern/Airtours as G-PIDS	N505EA	20.5.83
007	22196	225	Eastern/US Air as N602AU	N506EA	22.12.82
008	22197	225	Eastern/Airtours as G-RJGR	N507EA	28.12.82
009	22172	236	British Airways	G-BIKA	28.3.83
010	22173	236	British Airways	G-BIKB	25.1.83
011	22174	236	British Airways	G-BIKC	31.1.83
012	22198	225	Eastern/US Air as N603AU	N508EA	18.2.83
013	22175	236	British Airways	G-BIKD	10.3.83
014	22176	236	Air Europe/Star Air as N261PW	G-BKRM	30.3.83
015	22780	2T7	Monarch	G-MONB	22.3.83
016	22177	236	British Airways	G-BIKF	28.4.83
017	22199	225	Eastern/US Air as N604AU	N509EA	15.4.83
018	22781	2T7	Monarch	G-MONC	25.4.83
019	22960	2T7	Monarch	G-MOND	16.5.83
020	22200	225	Eastern/Airtours as G-MCEA	N510EA	28.6.83
021	22201	225	Eastern/US Air as N605AU	N511EA	28.7.83
022	22202	225	Eastern/US Air as N606AU	N512EA	19.8.83
023	22178	236	British Airways	G-BIKG	26.8.83
024	22179	236	British Airways	G-BIKH	18.10.83
025	22180	236	British Airways	G-BIKI	30.11.83
026	22203	225	Eastern/US Air as N607AU	N513EA	9.11.83
027	22204	225	Eastern/US Air as N608AU	N514EA	14.11.83
028	22205	225	Eastern/US Air as N609AU	N515EA	14.12.83
029	22181	236	British Airways	G-BIKJ	9.1.84
030	22182	236	British Airways	G-BIKK	1.2.84
031	22206	225	Eastern/Birgenair as TC-GEN, *crashed 6.2.96*	N516EA	26.2.85
032	22183	236	British Airways	G-BIKL	29.2.84
033	22184	236	British Airways	G-BIKM	21.3.84
034	22185	236	Air Europe/Iberia as EC-GCA	G-BPGW	27.3.84
035	22207	225	Eastern/AmericanWest as N913AW	N517EA	29.10.84
036	23118	2G5	LTU/LTE as EC-EFX	D-AMUR	25.5.84
037	22808	232	Delta	N601DL	28.2.85
038	22208	225	Eastern/AmericanWest as N914AW	N518EA	30.10.84

PRODUCTION HISTORY

L/N	C/N	Model	Operator (Original & Current)	Identity	Delivered
039	22809	232	Delta	N602DL	5.11.84
040	22209	225	Eastern/AmericanWest as N915AW	N519EA	21.11.84
041	22810	232	Delta	N603DL	7.11.84
042	22210	225	Eastern/US Air as N618AU	N520EA	30.11.84
043	22811	232	Delta	N604DL	3.12.94
044	23125	212	SIA/ATA as N751AT, Delta	9V-SGK	12.11.84
045	23126	212	SIA/ATA as N750AT, Delta	9V-SGL	26.11.84
046	22812	232	Delta	N605DL	7.12.84
047	23127	212	SIA/ATA as N757AT, Delta	9V-SGM	11.12.84
048	23128	212	SIA/ATA as N752AT, Delta	9V-SGN	12.12.84
049	22813	232	Delta	N606DL	17.1.85
050	22186	236	British Airways	G-BIKN	23.1.86
051	23119	2G5	LTU Sud/LTE as EC-EGH	D-AMUS	11.2.85
052	22187	236	British Airways	G-BIKO	14.2.85
053	23190	251	Northwest	N501US	28.2.85
054	22188	236	British Airways	G-BIKP	11.3.85
055	23191	251	Northwest	N502US	11.3.85
056	23293	2T7	Monarch	G-MONE	13.3.85
057	23227	236	Air Europe/Air Europa as EC-GCB	G-BLVH	26.3.85
058	22189	236	British Airways	G-BIKR	29.3.85
059	23192	251	Northwest	N503US	22.4.85
060	23193	251	Northwest	N504US	25.4.85
061	22814	232	Delta	N607DL	14.5.85
062	23194	251	Northwest	N505US	17.5.85
063	22190	236	British Airways	G-BIKS	31.5.85
064	22815	232	Delta	N608DA	31.5.85
065	22816	232	Delta	N609DL	11.6.85
066	22817	232	Delta	N610DL	28.6.85
067	23195	251	Northwest	N506US	8.7.85
068	23196	251	Norhwest	N507US	22.7.85
069	23197	251	Northwest	N508US	23.8.85
070	23198	251	Northwest	N509US	4.10.85
071	22818	232	Delta	N611DL	23.8.85
072	23199	251	Northwest	N511US	22.10.85

Line No. 74: Air 2000 757-225 G-OOOV (c/n 22211) was originally delivered to Eastern as N521EA on 6 December 1985 and worked until the airline was grounded in March 1991. After a period of storage the aircraft joined Air 2000 in October 1991 and it is seen on approach to Rhodes Airport in September 1995. *Philip Birtles*

BOEING 757

L/N	C/N	Model	Operator (original & current)	Identity	Delivered
073	22819	232	Delta	N612DL	18.10.85
074	22211	225	Eastern/Air 2000 as G-OOOV	N521EA	6.12.85
075	22611	225	Eastern/Air 2000 as G-OOOW	N522EA	5.12.85
076	23321	2S7	Republic/AmericanWest as N901AW	N601RC	19.12.85
077	23398	236	British Airways	G-BIKT	1.11.85
078	23399	236	British Airways	G-BIKU	7.11.85
079	23322	2S7	Republic/AWA as N902AW	N602RC	6.12.85
080	23323	2S7	Republic/AWA as N903AW	N603RC	30.12.85
081	23400	236	British Airways	G-BIKV	9.12.85
082	23200	251	Northwest	N512US	18.12.85
083	23201	251	Northwest	N513US	11.2.86
084	22820	232	Delta	N613DL	24.1.86
085	22821	232	Delta	N614DL	27.1.86
086	23202	251	Northwest	N514US	19.2.86
087	22822	232	Delta	N615DL	28.2.86
088	23203	251	Northwest	N515US	1.5.86
089	23492	236	British Airways	G-BIKW	7.3.86
090	23493	236	British Airways	G-BIKX	14.3.86
091	22823	232	Delta	N616DL	2.4.86
092	22907	232	Delta	N617DL	21.5.86
093	23533	236	British Airways	G-BIKY	28.3.86
094	23452	2M6	Royal Brunei	V8-RBA	6.5.86
095	22908	232	Delta	N618DL	25.4.86
096	23566	2S7	Republic/AWA as N904AW	N604RC	19.5.86
097	23567	2S7	Republic/AWA as N905AW	N605RC	19.5.86
098	23532	236	British Airways	G-BIKZ	15.5.86
099	23568	2S7	Republic/AWA as N906AW	N606RC	28.5.86
100	23453	2M6	Royal Brunei	V8-RBB	13.6.86
101	22909	232	Delta	N619DL	11.6.86
102	23454	2M6	Royal Brunei/Kazakhstan as P4-NSN	V8-RBC	29.7.86
103	23686	2B6	Royal Air Maroc	CN-RMT	15.7.86
104	23204	251	Northwest	N516US	8.7.86
105	23205	251	Northwest	N517US	1.8.86
106	23687	2B6	Royal Air Maroc	CN-RMZ	7.8.86
107	23206	251	Northwest	N518US	22.8.86
108	23207	251	Northwest	N519US	6.10.86
109	23208	251	Northwest	N520US	27.10.86
110	23209	251	Northwest	N521US	1.12.86
111	22910	232	Delta	N620DL	5.11.86
112	22911	232	Delta	N621DL	14.11.86
113	22912	232	Delta	N622DL	3.12.86
114	22612	225	Eastern/Air 2000 as G-OOOM	N523EA	11.11.86
115	22688	225	Eastern/LTU Sud as D-AMUU	N524EA	19.12.86
116	23651	2G5	LTU Sud/LTE as EC-ENQ	D-AMUT	8.12.86
117	22689	225	Eastern/LTU Sud as D-AMUK	N525EA	19.12.86
118	22913	232	Delta	N623DL	8.1.87
119	23616	251	Northwest	N522US	14.1.87
120	22914	232	Delta	N624DL	23.1.87
121	23617	251	Northwest	N523US	29.1.87
122	23618	251	Northwest	N524US	14.2.87
123	23710	236	British Airways	G-BMRA	2.3.87
124	23619	251	Northwest	N525US	14.4.87
125	23770	2T7	Monarch	G-DAJB	16.3.87
126	22915	232	Delta	N625DL	1.4.87
127	23767	28A	Air 2000/Canada 3000 as C-FOOA	G-OOOA	1.4.87
128	22916	232	Delta	N626DL	22.5.87
129	22917	232	Delta	N627DL	24.4.87
130	23822	28A	Air 2000/Canada 3000 as C-FOOB	G-OOOB	27.4.87
131	23620	251	Northwest	N526US	13.5.87
132	23895	2T7	Monarch/National as N513NA	G-DRJC	11.5.87
133	22918	232	Delta	N628DL	3.6.87
134	22919	232	Delta	N629DL	14.10.87

PRODUCTION HISTORY

L/N	C/N	Model	Operator (original & current)	Identity	Delivered
135	22920	232	Delta	N630DL	25.11.87
136	23842	251	Northwest	N527US	2.7.87
137	23843	251	Northwest	N528US	21.9.87
138	23612	232	Delta	N631DL	3.12.87
139	23723	24APF	UPS	N401UP	6.10.87
140	23844	251	Northwest	N529US	28.9.87
141	23724	24APF	UPS	N402UP	17.9.87
142	23850	2F8	Royal Nepal	9N-ACA	9.9.87
143	23725	24APF	UPS	N403UP	18.9.87
144	24014	21B	China Southern	B-2801	22.9.87
145	23975	236	British Airways	G-BMRB	25.9.87
146	23928	2G5	LTU Sud	D-AMUV	1.10.87
147	23726	24APF	UPS	N404UP	27.10.87
148	24015	21B	China Southern	B-2802	27.10.87
149	23727	24APF	UPS	N405UP	29.10.87
150	24016	21B	China Southern	B-2803	7.11.87
151	22690	225	Mexican AF TP-01	XC-CBD	16.11.87
152	23917	258	El Al	4X-EBL	25.11.87
153	23929	2G5	LTU Sud	D-AMUW	18.11.87
154	23613	232	Delta	N632DL	9.12.87
155	22691	225	American West	N907AW	10.12.87
156	23918	258	El Al	4X-EBM	17.12.87
157	23614	232	Delta	N633DL	23.12.87
158	23615	232	Delta	N634DL	13.1.88
159	23762	232	Delta	N635DL	3.2.88
160	24072	236	British Airways	G-BMRC	22.1.88
161	23983	2G5	LTU Sud	D-AMUX	8.3.88
162	24017	28A	Air 2000/Canada 3000 as C-FXOC	G-OOOC	10.3.88
163	24118	236	Air Europe/Royal as C-GRYO	G-BNSD	1.3.88
164	23763	232	Delta	N636DL	3.3.88
165	24135	27B	Air Holland	PH-AHE	9.3.88
166	24073	236	British Airways	G-BMRD	29.2.88
167	24119	236	Air Europa/Royal as C-GRYZ	EC-EHY	5.4.88
168	24074	236	British Airways	G-BMRE	23.3.88
169	24136	27B	Air Holland/Arkia as 4X-EBF	PH-AHF	23.3.88
170	24104	2T7	Monarch	G-MONJ	22.4.88

Line No. 130: Air 2000 757-23A G-OOOB (c/n 23822) was originally delivered on 27 April 1987, and leased to BA from 1 November 1987 until 23 April 1988. It has also operated with Canada 3000 as C-FOOB on a number of occasions. It is seen here while leased to BA on approach to London Heathrow in February 1988. *Philip Birtles*

BOEING 757

L/N	C/N	Model	Operator (original & current)	Identity	Delivered
171	23760	232	Delta	N637DL	13.4.88
172	24105	2T7	Monarch	G-MONK	15.4.88
173	24176	2G5	LTU Sud	D-AMUY	21.4.88
174	24120	236	Air Europe/BA as G-BPEF	G-BOHC	4.5.88
175	24101	236	British Airways	G-BMRF	13.5.88
176	23728	24APF	UPS	N406UP	2.6.88
177	23761	232	Delta	N638DL	25.5.88
178	24137	27B	Air Holland	PH-AHI	31.5.88
179	24102	236	British Airways	G-BMRG	31.5.88
180	24235	28A	Air 2000/Canadaa 3000 as C-FXOD	G-OOOD	27.5.88
181	23729	24APF	UPS	N407UP	24.6.88
182	23863	2F8C	Royal Nepal	9N-ACB	15.9.88
183	24121	236	Air Europe/Arkia as 4X-BAZ	G-BNSE	30.6.88
184	23730	24APF	UPS	N408UP	15.7.88
185	24254	258	El Al	4X-EBR	19.7.88
186	23731	24APF	UPS	N409UP	29.7.88
187	24122	236	Air Europe/Air Europa as EC-FFK	G-BNSF	29.7.88
188	23845	251	Northwest	N530US	10.8.88
189	23732	24APF	UPS	N410UP	1.9.88
190	23846	251	Northwest	N531US	25.8.88
191	23851	24APF	UPS	N411UP	1.9.88
192	24263	251	Northwest	N532US	27.9.88
193	23852	24APF	UPS	N412UP	30.9.88
194	24264	251	Northwest	N533US	30.9.88
195	23853	24APF	UPS	N413UP	30.9.88
196	24265	251	Northwest	N534US	21.10.88
197	23854	24APF	UPS	N414UP	28.10.88
198	23993	232	Delta	N639DL	9.11.88
199	23855	24APF	UPS	N415UP	28.10.88
200	24330	21B	China Southern	B-2804	22.11.88
201	23994	232	Delta	N640DL	30.11.88
202	23995	232	Delta	N641DL	7.12.88
203	24331	21B	China Southern	B-2805	16.12.88
204	24260	28A	Odyssey/National as NN517NA	C-FNBC	20.12.88
205	23996	232	Delta	N642DL	22.12.88
206	23997	232	Delta	N643DL	13.1.89
207	23998	232	Delta	N644DL	18.1.89
208	24367	28A	Odyssey/Flying Colours as G-FCLG	C-GAWB	1.2.89
209	24289	23A	Hispania Lineas/Air 2000 as G-OOOI	EC-EMV	15.2.89
210	24266	236	British Airways	G-BMRH	21.2.89
211	24267	236	British Airways	G-BMRI	17.2.89
212	24290	23A	Hispania Lineas/Air 2000 as G-OOOJ	EC-EMU	27.2.89
213	24368	28A	Monarch/ATA as N521AT	G-MCKE	3.3.89
214	24268	236	British Airways	G-BMRJ	6.3.89
215	24291	23A	Air Holland	PH-AHK	2.3.89
216	24216	232	Delta	N645DL	6.4.89
217	24217	232	Delta	N646DL	5.4.89
218	24370	236	BA/Caledonian	G-BPEA	31.3.89
219	24292	23A	Air 2000/Canada 3000 as C-FOOG	G-OOOG	29.3.89
220	24293	23A	Air 2000/Canada 3000 as C-GOOH	G-OOOH	6.4.89
221	24397	236	Air Europe/Air 2000 as G-OOOS	G-BRJD	13.4.89
222	24218	232	Delta	N647DL	3.5.89
223	24372	232	Delta	N648DL	4.5.89
224	24398	236	Air Europe/BA as G-CPEL	G-BRJE	26.4.89
225	24371	236	BA/Caledonian	G-BPEB	27.4.89
226	24369	28A	Canada 3000	C-FOOE	5.5.89
227	24451	2G5	LTU Sud	D-AMUM	5.5.89
228	24497	2G5	LTU Sud	D-AMUZ	11.5.89
229	24389	232	Delta	N649DL	25.5.89
230	24390	232	Delta	N650DL	28.6.89
231	24471	26D	Shanghai Airlines	B-2808	4.8.89
232	24401	21B	China Southern	B-2806	28.8.89

PRODUCTION HISTORY

L/N	C/N	Model	Operator (original & current)	Identity	Delivered
233	24402	21B	China Southern	B-2807	28.8.89
234	24486	223	American	N610AA	31.7.89
235	24472	26D	Shanghai Airlines	B-2809	29.8.89
236	24487	223	American	N611AM	17.7.89
237	24456	23APF	Challenge/Icelandair as TF-FIG	N571CA	26.7.89
238	24391	232	Delta	N651DL	26.7.89
239	24392	232	Delta	N652DL	27.7.89
240	24488	223	American	N612AA	4.8.89
241	24622	222	United	N501UA	24.8.89
242	24489	223	American	N613AA	11.8.89
243	24490	223	American	N614AA	18.8.89
244	24233	2G7	American West	N908AW	25.8.89
245	24491	223	American	N615AM	12.8.89
246	24623	222	United	N502UA	14.9.89
247	24624	222	United	N503UA	20.9.89
248	24524	223	American	N616AA	28.9.89
249	24527	23A	Petrolair	HB-IEE	2.10.89
250	24528	23A	Air Belgium/Air Holland as PH-AHP	OO-ILI	6.10.89
251	24625	222	United	N504UA	16.10.89
252	24522	2G7	American West	N909AW	7.11.89
253	24525	223	American	N617AM	17.11.89
254	14626	222	United	N505UA	13.11.89
255	24566	23A	Kenya Airways/TAESA as XA-RLM	5Y-BGI	6.12.89
256	24523	2G7	American West	N910AW	29.11.89
257	24567	23A	North American Airlines	N757NA	1.3.90
258	24635	23APF	Zambia Airways/DHL as VH-AWE	9J-AFO	18.10.90
259	24636	23A	Inter European/Airtours as G-LCRC	G-IEAB	1.2.90
260	24526	223	American	N618AA	5.2.90
261	24393	232	Delta	N653DL	9.2.90
262	24714	21B	China Southern	B-2811	15.2.90
263	24627	222	United	N506UA	21.2.90
264	24394	232	Delta	N654DL	22.2.90

Line No. 160: BA 757-236 G-BMRC (c/n 24072) was delivered on 22 January 1988 and is seen on approach to London Heathrow in September 1999 with the fin in Olympic team colours ready for Australia 2000. *Philip Birtles*

BOEING 757

L/N	C/N	Model	Operator (original & current)	Identity	Delivered
265	24395	232	Delta	N655DL	28.2.90
266	24396	232	Delta	N656DL	7.3.90
267	24737	230	Condor	D-ABNA	19.3.90
268	24543	28A	Odyssey/Air Transat as C-GTSN	C-GTDL	12.3.90
269	24577	223	American	N619AA	19.3.90
270	24743	222	United	N507UA	20.3.90
271	24772	236	Air Europe/Air Transat as C-GTSJ	G-BRJF	26.3.90
272	24771	236	Air Europe/National as N506NA	G-BRJG	27.3.90
273	24739	208	Icelandair	TF-FIH	4.4.90
274	24738	230	Condor	D-ABNB	5.4.90
275	24747	230	Condor	D-ABNC	11.4.90
276	24578	223	American	N620AA	17.4.90
277	24744	222	United	N508UA	18.4.90
278	24794	236	Air Europe/Iberia as EC-HDG	G-BRJH	20.4.90
279	24792	236	Air Europe/Britannia AB as SE-DUO	G-BRJI	30.4.90
280	24544	28A	Canada 3000	C-FXOF	30.4.90
281	24760	208	Icelandair	TF-FII	3.5.90
282	24758	21B	China Southern,	B-2812	16.5.90
			w/o 2.10.90 at Guangzhou when hit by landing 737		
283	24579	223	American	N621AM	17.5.90
284	24763	222	United	N509UA	21.5.90
285	24748	230	Condor	D-ABND	24.5.90
286	24419	232	Delta	N657DL	21.5.90
287	24420	232	Delta	N658DL	24.5.90
288	24774	21B	China Southern	B-2815	5.6.90
289	24580	223	American	N622AA	11.6.90
290	24780	222	United	N510UA	13.6.90
291	24799	222	United	N511UA	20.6.90
292	24793	236	Air Europe/Britannia AB as SE-DUP	G-BRJJ	20.6.90
293	24421	232	Delta	N659DL	25.6.90
294	24422	232	Delta	N660DL	22.6.90
295	24749	230	Condor	D-ABNE	12.7.90
296	24581	223	American	N623AA	11.7.90
297	24582	223	American	N624AA	16.7.90
298	24809	222	United	N512UA	20.7.90
299	24810	222	United	N513UA	24.7.90
300	24845	260PF	Ethiopian	ET-AJS	24.8.90
301	24473	26D	Shanghai Airlines	B-2810	3.8.90
302	24838	27B	Sterling Awys/Condor as D-ABNX	PH-AHL	2.8.90
303	24583	223	American	N625AA	8.8.90
304	24584	223	American	N626AA	14.8.90
305	24839	222	United	N514UA	17.8.90
306	24840	222	United	N515UA	21.8.90
307	24860	222	United	N516UA	24.8.90
308	24585	223	American	N627AA	10.9.90
309	24586	223	American	N628AA	31.8.90
310	24861	222	United	N517UA	24.9.90
311	24871	222	United	N518UA	14.9.90
312	24872	222	United	N519UA	14.9.90
313	24890	222	United	N520UA	21.9.90
314	24868	23APF	Challenge Air Cargo	N572CA	24.9.90
315	24587	223	American	N629AA	28.9.90
316	24588	223	American	N630AA	4.10.90
317	24589	223	American	N631AA	9.10.90
318	23903	24APF	UPS	N416UP	11.10.90
319	24891	222	United	N521UA	16.10.90
320	24931	222	United	N522UA	19.10.90
321	24590	223	American	N632AA	26.10.90
322	23904	24APF	UPS	N417UP	26.10.90
323	24882	236	British Airways/Caledonian	G-BPEC	6.11.90
324	24591	223	American	N633AA	16.11.90
325	24884	258	El Al	4X-EBS	13.11.90

PRODUCTION HISTORY

L/N	C/N	Model	Operator (Original & Current)	Identity	Delivered
326	23905	24APF	UPS	N418UP	14.11.90
327	24592	223	American	N634AA	19.11.90
328	24593	223	American	N635AA	28.11.90
329	24932	222	United	N523UA	29.11.90
330	23906	24APF	UPS	N419UP	4.12.90
331	24977	222	United	N524UA	10.12.90
332	24923	23A	Freeport McMoran/Vulcan NW	N680FM	30.5.91
333	24924	23A	Canada 3000	C-FXOK	12.3.91
334	23907	24APF	UPS	N420UP	20.12.90
335	24972	232	Delta	N661DN	23.12.90
336	24594	223	American	N636AM	18.1.91
337	24595	223	American	N637AM	16.1.91
338	24978	222	United	N525UA	15.1.91
339	24994	222	United	N526UA	22.1.91
340	24971	23APF	Anglo Cargo/Challenge as N573CA	G-OBOZ	29.8.91
341	24995	222	United	N527UA	27.2.91
342	24991	232	Delta	N662DN	28.1.91
343	24992	232	Delta	N663DN	9.2.91
344	24596	223	American	N638AA	22.2.91
345	24597	223	American	N639AA	26.2.91
346	25018	222	United	N528UA	19.2.91
347	25012	232	Delta	N664DN	21.2.91
348	25014	260	Ethiopian	ET-AJX	25.2.91
349	25013	232	Delta	N665DN	1.3.91
350	24598	223	American	N640A	15.3.91
351	24599	223	American	N641AA	25.3.91
352	25019	222	United	N529UA	22.3.91
353	25043	222	United	N530UA	20.3.91
354	25034	232	Delta	N666DN	22.3.91
355	25035	232	Delta	N667DN	27.3.91
356	25036	258	El Al	4X-EBT	1.4.91
357	24600	223	American	N642AA	5.4.91
358	25053	236	Air Europa	EC-FEE	11.4.91
359	25083	21B	China Southern	B-2816	14.4.91
360	24601	223	American	N643AA	19.4.91
361	25042	222	United	N531UA	19.4.91

Line No. 250: Air Belgium 757-23A OO-ILI (c/n 24528) was delivered on 6 October 1989 and is seen at Rhodes Airport in September 1995. It was leased to Sunways as SE-DSM in March 1996, until they ceased flying in November 1997. The aircraft then joined Britannia Airways as G-BXOL, moving to Air Holland Charter in April 1999 as PH-AHP. *Philip Birtles*

BOEING 757

L/N	C/N	Model	Operator (original & current)	Identity	Delivered
362	25054	236	TAESA/Britannia AB as SE-DUK	XA-MMX	4.12.91
363	25059	236	British Airways	G-BPED	30.4.91
364	25060	236	British Airways/Caledonian	G-BPEE	3.5.91
365	24602	223	American	N644AA	8.5.91
366	25072	222	United	N532UA	15.5.91
367	25073	222	United	N533UA	16.5.91
368	25085	208	Icelandair/Britannia as G-BTEJ	TF-FIJ	23.5.91
369	25044	2Q8	LADECO/FEAT as B27001	CC-CYG	28.5.91
370	24603	223	American	N645AA	30.5.91
371	25155	2J4	Sterling/Vulcan NW as N757AF	OY-SHA	7.6.91
372	25129	222	United	N534UA	7.6.91
373	25130	222	United	N535UA	12.6.91
374	25133	236	Air EuropeSpA/Mexicana asXA-TJC	I-AEJA	18.6.91
375	24604	223	American	N646AA	21.6.91
376	25141	232	Delta	N668DN	26.6.91
377	25142	232	Delta	N669DN	28.6.91
378	24605	223	American	N647AA	2.7.91
379	24606	223	American	N648AA	10.7.91
380	25156	222	United	N536UA	31.7.91
381	25157	222	United	N537UA	18.7.91
382	25140	230	Condor	D-ABNF	29.7.91
383	24607	223	American	N649AA	25.7.91
384	24608	223	American	N650AA	31.7.91
385	25222	222	United	N538UA	2.8.91
386	25223	222	United	N539UA	7.8.91
387	25220	2J4	Sterling/Yucaipa as N770BB	OY-SHB	25.9.91
388	25240	2Y0	Air 2000	G-OOOU	30.8.91
389	25258	21B	China Southern	B-2817	22.8.91
390	24609	223	American	N651AA	27.8.91
391	24610	223	American	N652AA	28.8.91
392	25259	21B	China Southern	B-2818	3.9.91
393	25252	222	United	N540UA	23.9.91
394	25253	222	United	N541UA	27.9.91
395	25281	24APF	UPS	N421UP	27.9.91
396	25276	222	United	N542UA	4.10.91
397	24611	223	American	N653A	8.10.91
398	24612	223	American	N654A	10.10.91
399	25324	24APF	UPS	N422UP	25.10.91
400	25268	2Y0	TAESA/BA as G-CPEP	XA-TAE	19.10.91
401	25698	22	United	N543UA	29.10.91
402	24613	223	American	N655AA	29.10.91
403	25325	24APF	UPS	N423UP	31.10.91
404	24614	223	American	N656AA	6.11.91
405	25322	222	United	N544UA	10.11.91
406	25323	222	United	N545UA	13.11.91
407	25369	24APF	UPS	N424UP	15.11.91
408	25353	260	Ethiopian	ET-AKC	18.11.91
409	24615	223	American	N657AM	6.12.91
410	24616	223	American	N658AA	27.11.91
411	25370	24APF	UPS	N425UP	5.12.91
412	25345	23A	Government of Turkmenistan	EZ-A010	1.12.93
413	25367	222	United	N546UA	13.12.91
414	25368	222	United	N547UA	20.12.91
415	25331	232	Delta	N670DN	9.1.92
416	25332	232	Delta	N671DN	10.1.92
417	24617	223	American	N659AA	9.1.92
418	25294	223	American	N660AM	17.1.92
419	25436	230	Condor	D-ABNH	29.1.92
420	25396	222	United	N548UA	30.1.92
421	25397	222	United	N549UA	21.1.92
422	25437	230	Condor	D-ABNI	30.1.92
423	25295	223	American	N661AA	30.1.92

PRODUCTION HISTORY

L/N	C/N	Model	Operator (original & current)	Identity	Delivered
424	24964	2Q8	ATA/Mexicana as N755MX	N754AT	5.2.92
425	25296	223	American	N662AA	7.2.92
426	25398	222	United	N550UA	11.2.92
427	25399	222	United	N551UA	18.2.92
428	25438	230	Condor	D-ABNK	20.2.92
429	25977	232	Delta	N672DL	21.2.92
430	25978	232	Delta	N673DL	24.2.92
431	26641	222	United	N552UA	27.2.92
432	25297	223	American	N663AM	5.3.92
433	25298	223	American	N664AA	2.3.92
434	25277	222	United	N553UA	10.3.92
435	26644	222	United	N554UA	12.3.92
436	25299	223	American	N665AA	17.3.92
437	25439	230	Condor	D-ABNL	19.3.92
438	24965	2Q8	ATA/Mexicana as N758MX	N755AT	24.3.92
439	25979	232	Delta	N674DL	25.3.92
440	26962	204	Britannia/National as N512NA	G-BYAC	10.4.92
441	25597	236	LTE/Air Europa as EC-CBX	EC-FLY	15.4.92
442	26647	222	United	N555UA	16.4.92
443	25440	230	Condor	D-ABNM	9.4.92
444	26057	260	Ethiopian	ET-AKE	22.4.92
445	25598	236	China Southern	B-2835	3.6.93
446	25441	230	Condor	D-ABNN	21.4.92
447	26650	222	United	N556UA	23.4.92
448	25980	232	Delta	N675DL	29.4.92
449	25620	236	Inter Eur./Greenlandair as OY-GRL	G-IEAC	28.4.92
450	26963	204	Britannia	G-BYAD	6.5.92
451	25300	223	American	N666A	7.5.92
452	26964	204	Britannia	G-BYAE	12.5.92
453	25592	236	Caledonian/National as N521NA	G-BUDX	13.5.92
454	26653	222	United	N557UA	19.5.92
455	25981	232	Delta	N676DL	28.5.92
456	25982	232	Delta	N677DL	22.5.92
457	25621	28A	Canada 3000	C-FXOO	18.5.92
458	25131	2Q8	LADECO/LAPA as LV-WTS	CC-CYH	2.6.92
459	25301	223	American	N7667A	4.6.92
460	25333	223	American	N668AA	10.6.92

Line No. 374: Sunways Airlines, a Swedish charter operator, were only in operation for 11 months from October 1996. Boeing 757-236 SE-DSN (c/n 25133) was originally delivered to Air Europa as I-AEJA, before becoming OO-TBI with BFS and B-17501 with Makung. It is seen turning on to the runway at Rhodes Airport in August 1997 and it is now XA-TJC with Mexicana. *Philip Birtles*

BOEING 757

L/N	C/N	Model	Operator (original & current)	Identity	Delivered
461	25884	21B	China Southern	B-2822	11.6.92
462	26654	222	United	N558UA	17.6.92
463	25334	223	American	N669AA	24.6.92
464	25901	230	Condor	D-ABNO	27.6.92
465	25983	232	Delta	N678DL	26.6.92
466	25593	236	Caledonian/Royal as C-GRYK	G-BUDZ	25.6.92
467	26657	222	United	N559UA	6.7.92
468	25335	223	American	N670AA	10.7.92
469	26660	222	United	N560UA	13.7.92
470	25487	23A	Argentine Government	T-01	22.7.92
471	25488	23A	Air Transat	C-GTSE	20.11.92
472	26151	2Y0	TAESA/Britannia AB as SE-DUL	XA-KWK	23.7.92
473	25336	223	American	N671AA	24.7.92
474	25337	223	American	N672AA	29.7.92
475	25898	25C	Xiamen Airlines	B-2819	12.8.92
476	25885	21B	China Southwest	B-2820	5.8.92
477	25457	24APF	UPS	N426UP	13.8.92
478	26152	2Y0	Avianca	EI-CEY	13.8.92
479	26661	222	United	N561UA	17.8.92
480	25886	21B	China Southwest	B-2821	19.8.92
481	25458	24APF	UPS	N427UP	3.9.92
482	26153	2Y0	China SW/Xinjiang	B-2831	26.8.92
483	25338	223	American	N681AA	28.8.92
484	25339	223	American	N682AA	2.9.92
485	25459	24APF	UPS	N428UP	10.9.92
486	26154	2Y0	Avianca	EI-CEZ	22.9.92
487	26664	222	United	N562UA	28.9.92
488	26665	222	United	N563UA	29.9.92
489	25460	24APF	UPS	N429UP	8.10.92
490	26666	222	United	N564UA	8.10.92
491	25340	223	American	N683A	14.10.92
492	26669	222	United	N565UA	15.10.92
493	25461	24APF	UPS	N430UP	22.10.92
494	26670	222	United	N566UA	23.10.92
495	26155	2Y0	China SW	B-2826	29.10.92
496	26058	260	Ethiopian	ET-AKF	29.10.92
497	26673	222	United	N567UA	2.11.92
498	26674	222	United	N568UA	5.11.92
499	26677	222	United	N569UA	10.11.92
500	26955	232	Delta	N679DA	12.11.92
501	26678	222	United	N570UA	19.11.92
502	26856	232	Delta	N680DA	19.11.92
503	26156	2Y0	China SW/Xinjiang	B-2827	25.11.92
504	25341	223	American	N684AA	2.12.92
505	25489	23A	Aeromextour/Aero Peru as N52AW lost off Lima 2.10.96	XA-SKR	27.9.93
506	26681	222	United	N571UA	7.12.92
507	25342	223	American	N685AA	9.12.92
508	26682	222	United	N572UA	14.12.92
509	25343	223	American	N686AA	16.12.92
510	25490	23A	Aeromextour/Aeromexico asN53AW	XA-SKQ	18.12.92
511	25491	23A	Air Transat	C-GTSF	10.12.92
512	26685	222	United	N573UA	22.12.92
513	26686	222	United	N574UA	23.12.92
514	26266	204	Britannia	G-BYAF	13.1.93
515	26689	222	United	N575UA	2.2.93
516	26957	232	Delta	N681DA	23.1.92
517	26865	204	Britannia, w/o Gerona, Spain 14.9.99	G-BYAG	22.1.93
518	26958	232	Delta	N682DA	15.1.93
519	26633	2K2	Transavia	PH-TKA	22.2.93
520	26966	204	Britannia	G-BYAH	5.2.93
521	26433	230	Condor	D-ABNP	12.2.93

PRODUCTION HISTORY

L/N	C/N	Model	Operator (original & current)	Identity	Delivered
522	26967	204	Britannia	G-BYAI	1.3.93
523	25493	23A	American Trans Air	N512AT	30.9.94
524	26690	222	United	N576UA	17.2.93
525	27122	2B7	US Air	N610AU	26.2.93
526	26158	2Y0	Air 2000	G-OOOX	24.2.93
527	26693	222	United	N577UA	1.3.93
528	25623	28A	Britannia	G-BYAJ	4.3.93
529	26053	258	El Al	4X-EBU	8.3.93
530	25622	28A	Air Seychelles/A. Transat as C-GTSV	S7-AAX	10.3.93
531	26694	222	United	N578UA	12.3.93
532	26434	230	Condor	D-ABNR	16.3.93
533	27103	232	Delta	N683DA	18.3.93
534	27123	2B7	US Air	N611AU	22.3.93
535	27104	232	Delta	N684DA	26.3.93
536	25695	223	American	N687AA	31.3.93
537	26435	230	Condor	D-ABNS	2.4.93
538	26267	204	Britannia	G-BYAK	6.4.93
539	26697	222	United	N579UA	9.4.93
540	27124	2B7	US Air	N612AU	12.4.93
541	25624	2Q8	Aeromexico	XA-SIK	16.4.93
542	26698	222	United	N580UA	21.4.93
543	26701	222	United	N581UA	3.5.93
544	27144	2B7	US Air	N613AU	27.4.93
545	26634	2K2	Transavia	PH-TKB	3.5.93
546	27145	2B7	US Air	N614AU	3.5.93
547	26054	258	El Al	4X-EBV	5.5.93
548	25730	223	American	N688AA	12.5.93
549	25626	204	Britannia	G-BYAL	13.5.93
550	26702	222	United	N582UA	19.5.93
551	27146	2B7	US Air	N615AU	19.5.93
552	27147	2B7	US Air	N616AU	24.5.93
553	26239	256	Iberia	EC-FTR	7.6.93
554	25887	2Z0	China Southwest	B-2832	4.6.93
555	26160	2Y0	Transaero/Flying Colours as G-FCLJ	EI-CJX	1.4.94
556	26705	222	United	N583UA	14.6.93
557	26161	2Y0	Transaero/Flying Colours asG-FCLK	EI-CJY	13.4.94
558	26270	2Q8	Aeromexico	XA-SJD	18.6.93
559	26706	222	United	N584UA	24.6.93

Line No. 449: Inter European 757-236 G-IEAC (c/n 25620) was delivered on 28 April 1992 on lease from Sunrock and operated holiday charter flights until it ceased flying in October 1993. The aircraft went to Airtours as G-CSVS until April 1998, when it joined Greenlandair first as TF-GRL and then OY-GRL. It is seen on approach to London Gatwick in July 1993. *Philip Birtles*

BOEING 757

L/N	C/N	Model	Operator (original & current)	Identity	Delivered
560	27152	26D	Shanghai Airlines	B-2833	29.6.93
561	26240	256	Iberia	EC-FUA	12.8.93
562	25731	223	American	N689AA	9.7.93
563	26709	222	United	N585UA	13.7.93
564	27148	2B7	US Air	N617AU	15.7.93
565	25899	25C	Xiamen Airlines	B-2828	21.7.93
566	25696	223	American	N690AA	26.7.93
567	26710	222	United	N586UA	29.7.93
568	25697	223	American	N691AA	11.8.93
569	25462	24APF	UPS	N431UP	12.8.93
570	26713	222	United	N587UA	11.8.93
571	26717	222	United	N588UA	16.8.93
572	26241	256	Iberia	EC-FUB	17.8.93
573	25463	24APF	UPS	N432UP	2.9.93
574	25900	25C	Xiamen	B-2829	30.8.93
575	25888	21B	China Southern	B-2823	17.9.93
576	27183	26D	Shanghai Airlines	B-2834	24.9.93
577	25464	24APF	UPS	N433UP	30.9.93
578	26972	223	American	N692AA	8.10.93
579	25465	24APF	UPS	N434UP	14.10.93
580	26973	223	American	N693AA	7.1.94
581	25466	24APF	UPS	N435UP	21.10.93
582	26974	223	American	N694AN	26.1.94
583	25889	21B	China Southern	B-2824	4.11.93
584	27198	2B7	US Air	N619AU	5.11.93
585	25890	21B	China Southern	B-2825	12.11.93
586	27199	2B7	US Air	N620AU	15.11.93
587	26436	230	Condor	D-ABNT	8.3.94
588	27203	29J	FEAT	B-27005	18.11.94
589	27200	2B7	US Air	N621AU	6.12.93
590	26268	2Q8	Aeromexico	XA-SMJ	7.1.94
591	27204	29J	FEAT	B-27007	8.11.94
592	26271	2Q8	Aeromexico	XA-SMK	27.1.94
593	26242	256	Iberia	EC-FYJ	27.7.94
594	26272	2Q8	Aeromexico	XA-SML	1.3.94
595	27258	2Z0	China SW	B-2836	25.2.94
596	27219	204	Britannia	G-BYAN	26.1.94
597	26273	2Q8	Aeromexico	XA-SMM	1.4.94
598	27235	204	Britannia	G-BYAO	3.2.94
599	25495	23A	Saudi Arabian Government	HZ-HMED	3.6.94
600	27236	204	Britannia	G-BYAP	15.2.94
601	25806	236	British Airways	G-BPEI	9.3.94
602	27237	204	Britannia	G-BYAR	1.3.94
603	26243	256	Iberia	EC-FYK	29.7.94
604	27238	204	Britannia	G-BYAS	9.3.94
605	27201	2B7	US Air	N622AU	14.3.94
606	27208	204	Britannia	G-BYAT	21.3.94
607	27244	2B7	US Air	N623AU	24.3.94
608	26635	2K2	Transavia	PH-TKC	12.4.94
609	27259	2Z0	China SW	B-2837	11.8.94
610	25807	236	British Airways	G-BPEJ	25.4.94
611	25494	23A	AVIANCA	N987AN	22.4.94
612	26269	2Q8	Baikal A/L/Avianca	N321LF	28.4.94
613	27260	2Z0	China SW	B-2838	2.5.94
614	27291	224	Continental	N58101	12.5.94
615	27269	2Z0	China SW	B-2839	12.8.94
616	26244	256	Iberia	EC-FYL	2.8.94
617	26245	256	Iberia	EC-FYM	3.8.94
618	27220	204	Britannia	G-BYAU	18.5.94
619	27292	224	Continental	N14102	3.6.94
620	26246	256	Iberia	EC-FYN	4.8.94
621	26975	223	American	N695AN	13.6.94

PRODUCTION HISTORY

L/N	C/N	Model	Operator (Original & Current)	Identity	Delivered
622	27270	2Z0	China SW	B-2840	12.8.94
623	27293	224	Continental	N33103	24.6.94
624	27367	2Z0	China SW	B-2841	15.8.94
625	25467	24APF	UPS	N436UP	7.7.94
626	27342	26D	Shanghai	B-2842	24.8.94
627	26976	223	American	N696AN	15.7.94
628	25468	24APF	UPS	N437UP	21.7.94
629	27294	224	Continental	N17104	29.7.94
630	27245	2B7	US Air	N624AU	29.7.94
631	25469	24APF	UPS	N438UP	11.8.94
632	27295	224	Continental	N17105	18.8.94
633	26977	223	American	N697AN	17.8.94
634	25470	24APF	UPS	N439UP	25.8.94
635	26980	223	American	N698AN	30.8.94
636	25471	24APF	UPS	N440UP	26.9.94
637	27296	224	Continental	N14106	22.9.94
638	27386	24APF	UPS	N441UP	29.9.94
639	27351	2Q8	ATA/Mexicana as N764MX	N756AT	4.10.94
640	27387	24APF	UPS	N442UP	13.10.94
641	27297	224	Continental	N14107	14.10.94
642	27388	24APF	UPS	N443UP	27.10.94
643	27246	2B7	US Air	N625VJ	26.10.94
644	27389	24APF	UPS	N444UP	3.11.94
645	27298	224	Continental	N21108	7.11.94
646	27390	24APF	UPS	N445UP	18.11.94
647	27303	2B7	US Air	N626AU	8.11.94
648	27299	224	Continental	N12109	5.12.94
649	27735	24APF	UPS	N446UP	2.12.94
650	27300	224	Continental	N13110	6.12.94
651	27736	24APF	UPS	N447UP	13.12.94
652	27301	224	Continental	N57111	14.12.94
653	27302	224	Continental	N18112	2.2.95
654	27737	24APF	UPS	N448UP	13.1.95
655	27805	2B7	US Air	N627AU	27.2.97
656	27738	24APF	UPS	N449UP	19.1.95
657	27806	2B7	US Air	N628AU	27.1.95
658	26277	28A	North American	N750NA	27.1.95

Line No. 784: Boeing 757-236 G-CPER (c/n 29113) was one of the last to be delivered to BA, when it was handed over on 29 December 1997. It is seen at London Gatwick on a rather damp July day in 1998 with the Wings tail design representing Denmark. *Philip Birtles*

BOEING 757

L/N	C/N	Model	Operator (original & current)	Identity	Delivered
659	25472	24APF	UPS	N450UP	9.2.95
660	27051	223	American	N699AN	10.2.95
661	27052	223	American	N601AN	17.2.95
662	27807	2B7	US Air	N629AU	24.2.95
663	27234	204	Britannia	G-BYAW	3.4.95
664	27053	223	American	N602AN	10.3.95
665	25808	236	British Airways	G-BPEK	17.3.95
666	27808	2B7	US Air	N630AU	24.3.95
667	27588	232	Delta	N685DA	31.3.95
668	27555	224	Continental	N13113	11.4.95
669	27511	2Z0	China SW	B-2844	4.5.95
670	27054	223	American	N603AA	21.4.95
671	26278	2G5	LTU	D-AMUQ	26.4.95
672	26275	28A	Transaero/Flying Colours as G-FCLI	EI-CLV	1.5.95
673	27809	2B7	US Air	N631AU	11.5.95
674	27512	2Z0	China SW	B-2845	1.6.95
675	27739	24APF	UPS	N451UP	1.6.95
676	26274	28A	Transaero/Flying Colours asG-FCLH	EI-CLU	6.6.95
677	27055	223	American	N604AA	12.6.95
678	27810	2B7	US Air	N632AU	27.2.97
679	25473	24APF	UPS	N452UP	5.7.95
680	27056	223	American	N605AA	28.6.95
681	27811	2B7	US Air	N633AU	7.7.95
682	27556	224	Continental	N12114	3.7.95
683	25474	24APF	UPS	N453UP	3.8.95
684	27681	26D	Shanghai Airlines	B-2843	26.7.95
685	27513	25C	Xiamen	B-2848	7.8.95
686	27557	224	Continental	N14115	14.8.95
687	25475	24APF	UPS	N454UP	8.9.95
688	26332	2Q8	LAPA	LV-WMH	15.9.95
689	27589	232	Delta	N686DA	20.9.95
690	27971	23N	American Trans Air	N514AT	26.9.95
691	25476	24APF	UPS	N455UP	28.9.95
692	27598	23N	American Trans Air	N515AT	11.10.95
693	26482	251	Northwest	N535US	14.11.95
694	27972	23N	American Trans Air	N516AT	4.12.95
695	26483	251	Northwest	N536US	11.12.95
696	27599	2Q8	Transportes Aereos de Cabo Verde	D4-CBG	15.3.96
697	26484	251	Northwest	N537US	20.2.96
698	27517	25C	Xiamen	B-2849	7.2.96
699	26485	251	Northwest	N538US	1.3.96
700	26486	251	Northwest	N539US	25.3.96
701	26487	251	Northwest	N540US	15.4.96
702	27558	224	Continental	N12116	27.3.96
703	26488	251	Northwest	N541US	19.4.96
704	26276	28A	Icelandair	TF-FIK	15.3.96
705	26489	251	Northwest	N542US	10.5.96
706	27559	224	Continental	N19117	24.4.96
707	27057	223	American	N606AA	11.4.96
708	28112	2G5	LTU	D-AMUI	14.4.96
709	26490	251	Northwest	N543US	15.5.96
710	26491	251	Northwest	N544US	20.5.96
711	26492	251	Northwest	N545US	20.6.96
712	27058	223	American	N607AM	16.5.96
713	26493	251	Northwest	N546US	19.7.96
714	26494	251	Northwest	N547US	23.8.96
715	26495	251	Northwest	N548US	30.8.96
716	26496	251	Northwest	N549US	20.9.96
717	26330	2K2	Transavia	PH-TKD	24.6.96
718	28142	222	United	N591UA	28.6.96
719	28143	222	United	N592UA	10.7.96
720	27446	223	American	N608AA	16.7.96

PRODUCTION HISTORY

L/N	C/N	Model	Operator (original & current)	Identity	Delivered
721	28160	2Q8	TWA	N701TW	22.7.96
722	27447	223	American	N609AA	29.7.96
723	28161	28A	Canada 3000	C-FOON	23.9.96
724	28144	222	United	N593UA	12.8.96
725	28336	22K	Turkmenistan	EZ-A011	29.8.96
726	28337	22K	Turkmenistan	EZ-A012	30.8.96
727	28145	222	United	N594UA	11.9.96
728	25477	24APF	UPS	N456UP	13.9.96
729	25478	24APF	UPS	N457UP	25.9.96
730	25479	24APF	UPS	N458UP	3.10.96
731	28338	23P	Uzbekistan	UK-75700	19.10.96
732	28162	2Q8	TWA	N702TW	22.10.96
733	25480	24APF	UPS	N459UP	31.10.96
734	25481	24APF	UPS	N460UP	14.11.96
735	27973	23N	American Trans Air	N517AT	25.11.96
736	27620	2Q8	TWA	N703TW	22.11.96
737	27974	23N	American Trans Air	N518AT	10.12.96
738	27621	28A	Flying Colours/jmc AIR	G-FCLA	26.2.97
739	28463	26D	Mid East Jet	N757MA	21.1.97
740	28446	26D	Delta	N900PC	21.1.97
741	28163	2Q8	TWA	N704X	30.1.97
742	28479	231	TWA	N705TW	10.2.97
743	28165	2Q8	TWA	N706TW	18.2.97
744	27625	2Q8	TWA	N707TW	24.2.97
745	27622	258	El Al	4X-EBI	24.3.97
746	28674	21K	Airtours	G-WJAN	19.3.97
747	28665	236	British Airways	G-CPEM	28.3.97
748	27560	224	Continental	N14118	21.3.97
749	28164	28A	Flying Colours/jmc AIR	G-FCLB	25.3.97
750	28480	231	TWA	N708TW	7.4.97
751	28666	236	British Airways	G-CPEN	23.4.97
752	28718	25F	Flying Colours/jmc AIR	G-FCLD	25.4.97
753	27561	224	Continental	N18119	12.5.97
754	28168	2Q8	TWA	N709TW	14.5.97
755	28265	24APF	UPS	N461UP	30.5.97
756	28166	2Q8	Flying Colours	G-FCLC	9.5.97
757	28169	2Q8	TWA	N710TW	29.5.97

Line Number 857: Finnair Boeing 757-2Q8 OH-LBU (c/n 29377) was the last of five aircraft to be delivered, arriving on 9 March 1999. Finnair operates its 757-200ERs on direct services from Helsinki to the Canaries, Middle East and India. *Finnair*

BOEING 757

L/N	C/N	Model	Operator (original & current)	Identity	Delivered
758	28481	231	TWA	N711ZX	3.6.97
759	28266	24APF	UPS	N462UP	19.6.97
760	27624	2Q8	TWA	N712TW	18.6.97
761	27562	224	Continental	N14120	24.6.97
762	28667	236	British Airways	G-CPEO	12.7.97
763	28267	24APF	UPS	N463UP	19.6.97
764	28173	2Q8	TWA	N713TW	16.7.97
765	28268	24APF	UPS	N464UP	9.9.97
766	27563	224	Continental	N14121	29.7.97
767	28269	24APF	UPS	N465UP	16.9.97
768	27564	224	Continental	N17122	12.8.97
769	25482	24APF	UPS	N466UP	30.9.97
770	28482	231	TWA	N714P	28.8.97
771	25483	24APF	UPS	N467UP	29.9.97
772	28172	2Q8	Finnair	OH-LBO	7.10.97
773	28707	222	United	N589UA	3.11.97
774	25484	24APF	UPS	N468UP	14.10.97
775	28167	2Q8	Finnair	OH-LBR	16.10.97
776	25485	24APF	UPS	N469UP	29.10.97
777	28483	231	TWA	N715TW	23.10.97
778	25486	24APF	UPS	N470UP	24.11.97
779	27975	23N	American Trans Air	N519AT	17.11.97
780	28989	208	Icelandair	TF-FIN	20.1.98
781	28966	224	Continental	N26123	12.12.97
782	28833	2Q8	China Xinjiang	B-2852	12.12.97
783	29025	2G4	USAF C-32B	98-0001	1.6.98
784	29113	236	British Airways	G-CPER	29.12.97
785	28708	222	United	N590UA	31.12.97
786	27565	224	Continental	N29124	16.1.98
787	29026	2G4	USAF C-32B	98-0002	29.5.98
788	28967	224	Continental	N12125	2.2.98
789	28748	222	United	N595UA	4.2.98
790	27566	224	Continental	N29124	16.1.98
791	28968	224	Continental	N48127	25.2.98
792	27623	2Q8	Finnair	OH-LBS	9.3.98

Line number 905: Iberia Boeing 757-256 EC-HDV (c/n 26254,) was one of a number of new 757-200s delivered to the Spanish carrier during 1999. It was handed over on 22 December 1999. It is seen here on approach to London Heathrow in February 2000. *Philip Birtles*

PRODUCTION HISTORY

Line number 81: British Airways 757-236 G-BIKV (c/n 23400) was delivered on 9 December 1985 and is seen here on approach to London Heathrow in September 1999. During that year BA adopted a revised livery based on the Union Jack scheme first applied to the Concorde fleet. Under current plans about half the fleet will be painted in this scheme over the coming years. BA currently operates a fleet of 53 757-200s, but with the first of these scheduled to leave in 2000 — for conversation to freighter status with DHL — the fleet will gradually be reduced to around 10 RB311-535-E4 powered 757s over the next few years. *Philip Birtles*

793	29114	236	British Airways	G-CPES	17.3.98
794	28749	222	United	N596UA	17.3.98
795	27567	224	Continental	N17128	20.3.98
796	28969	224	Continental	N29129	27.3.98
797	29215	28S	China Xinjiang	B-2851	29.4.98
798	29115	236	British Airways	G-CPET	12.5.98
799	28970	224	Continental	N19130	8.5.98
800	27586	232	Delta	N687DL	3.4.98
801	28170	2Q8	Finnair	OH-LBT	3.5.98
802	28203	2Q8	Air 2000	G-OOOY	21.5.98
803	27587	232	Delta	N688DL	21.5.98
804	29016	330	Condor	G-ABOA	25.6.99
805	28171	2Q8	Flying Colours	G-FCLE	24.6.98
806	28971	224	Continental	N34131	9.6.98
807	27172	232	Delta	N689DL	18.6.98
808	27585	232	Delta	N690DL	26.6.98
809	29281	224	Continental	N33132	30.6.98
810	29017	330	Condor	D-ABOB	20.5.99
811	29216	28S	China Xinjiang	B-2853	8.7.98
812	29423	223	American	N673AN	5.8.98
813	28842	24APF	UPS	N471UP	23.7.98
814	27976	23N	American Trans Air	N520AT	31.7.98
815	28843	24APF	UPS	N472UP	12.8.98
816	29424	223	American	N674AN	17.8.98
817	29425	223	American	N675AN	24.8.98
818	29015	330	Condor	D-ABOC	5.5.99
819	29442	2Q8	Mexicana	N762MX	12.9.98
820	29724	232	Delta	N692DL	4.9.98
821	29443	2Q8	Mexicana	N763MX	16.9.98
822	29792	2Z0	China SW/Royal Nepal	B-2855	28.9.98
823	28846	24APF	UPS	N473UP	22.9.98
824	29027	2G4	USAF C-32A	99-0003	20.11.98
825	28484	231	TWA	N716TW	15.10.98
826	29725	232	Delta	N693DL	17.10.98
827	29426	223	American	N676AN	27.10.98
828	29427	223	American	N677AN	2.11.98
829	29028	2G4	USAF C-32A	99-0004	25.11.98

BOEING 757

L/N	C/N	Model	Operator (original & current)	Identity	Delivered
830	29488	2G5	LTU	D-AMUG	31.10.98
831	29726	232	Delta	N694DL	11.11.98
832	29607	27A	FEAT	B-27011	8.12.98
833	29793	2Z0	China SW/Royal Nepal	B-2856	24.11.98
834	29489	2G5	LTU	D-AMUH	20.11.98
835	29608	27A	FEAT	B-27013	14.12.98
836	29380	2Q8ER	Aeromexico	N380RM	9.12.98
837	29428	223	American	N678AN	11.12.98
838	29727	232	Delta	N695DL	11.12.98
839	29012	330	Condor	D-ABOE	10.3.99
840	29282	224	Continental	N17133	22.12.98
841	28750	222	United	N597UA	11.1.99
842	29589	223	American	N679AN	15.1.99
843	29330	23N	American Trans Air	N522AT	30.12.98
844	28751	222	United	N598UA	22.1.99
845	29728	232	Delta	N696DL	27.1.99
846	29013	330	Condor	D-ABOF	13.3.99
847	29590	223	American	N680AN	28.1.99
848	29283	224	Continental	N67134	2.2.99
849	29014	330	Condor	D-ABOG	19.3.99
850	28834	204	Britannia	G-BYAX	24.2.99
851	29284	224	Continental	N41135	19.2.99
852	29591	223	American	N181AN	27.2.99
853	29592	223	American	N182AN	3.3.99
854	28485	231	TWA	N717TW	15.3.99
855	30031	330	Condor	D-ABOH	19.3.99
856	29285	224	Continental	N19136	19.3.99
857	29377	2Q8	Finnair	OH-LBU	9.3.99
858	28835	28A	Flying Colours/jmc AIR	G-FCLF	23.3.99
859	29436	208	Icelandair	TF-FIO	20.4.99
860	26247	256	Iberia	EC-GZY	30.4.99
861	28836	204	Britannia	G-BYAY	13.4.99
862	29593	223	American	N183AN	22.4.99
863	26248	256	Iberia	EC-GZZ	30.4.99
864	29941	236	British Airways	G-CPEU	1.5.99
865	28174	28A	North American Airlines	N752NA	12.5.99
866	29594	223	American	N184AN	13.5.99
867	29942	236	National	N544NA	14.6.99
868	29217	28S	China Xinjiang	B-2859	2.6.99
869	28486	231	TWA	N718TW	26.5.99
870	29304	22L	Starflight	N1018N	17.12.99
871	29943	236	British Airways	G-CPEV	11.6.99
872	29944	236	National	N545NA	14.6.99
873	29945	236	National	N546NA	21.6.99
874	29954	231	TWA	N721TW	29.6.99
875	30060	23P	Uzbekistan	VP-BUB	3.9.99
876	29609	27A	FEAT	B-27015	8.7.99
877	29946	236	National	N547NA	11.8.99
878	28487	231	TWA	N719TW	26.7.99
879	28844	24APF	UPS	N474UP	29.7.99
880	30318	232	Delta	N697DL	1.8.99
881	26249	256	Iberia	EC-HAA	16.8.99
882	28845	24APF	UPS	N475UP	12.8.99
883	30319	231	TWA	N720TW	16.8.99
884	28488	231	TWA	N724TW	8.9.99
885	29911	232	Delta	N698DL	31.8.99
886	30061	23P	Uzbekistan	VP-BUD	9.12.99
887	29970	232	Delta	N699DL	13.9.99
888	30232	23N	American Trans Air	N523AT	21.9.99
889	26250	256	Iberia	EC-HDM	22.9.99
890	30337	232	Delta	N6700	24.9.99
891	30338	231	TWA	N725TW	12.10.99

PRODUCTION HISTORY

L/N	C/N	Model	Operator (original & current)	Identity	Delivered
892	30187	232	Delta	N6701	7.10.99
893	29385	231	TWA	N722TW	25.10.99
894	29305	22L	Azerbaijan	4K-AZ12	
895	30233	23N	American Trans Air	N524AT	25.10.99
896	30339	231	TWA	N726TW	18.11.99
897	26251	256	Iberia	EC-HDR	12.11.99
898	30188	232	Delta	N6702	20.11.99
899	30229	224	Continental	N34137	29.11.99
900	26252	256	Iberia	EC-HDS	2.12.99
901	30340	231	TWA	N727TW	2.12.99
902	26253	256	Iberia	EC-HDU	7.12.99
903	30351	224	Continental	N13138	7.12.99
904	29610	27A	FEAT	B27017	22.12.99
905	26254	256	Iberia	EC-HDV	22.12.99
906	30178	3E7	Arkia	4X-BAU	31.1.00
907	29378	231	TWA	N723TW	18.1.00
908	30234	232	Delta	N6703D	21.1.00
909	29018	330	Condor	D-ABOI	
910	29611	27A	FEAT	B-27021	27.1.00
911	30352	224	Continental	N17139	3.2.00
912	30179	3E7	Arkia	4X-BAW	25.2.00
913	30353	224	Continental	N41140	28.2.00
914	30396	232	Delta	N6704Z	
915	29019	330	Condor	D-ABOJ	
916					
917					
918	29020	330	Condor	D-ABOK	
919					
920					
921					
922					
923	29021	330	Condor	D-ABOL	
924					
925					
926	29022	330	Condor	D-ABOM	

Line number 909: Condor 757-330 D-ABOI (c/n 29018) was taken on a European sales tour in early 2000, visiting Manchester and Luton Airports in the UK. Orders followed in the shape of two aircraft for Britain's jmc AIR announced on 23 March, for delivery in time for the 2001 holiday season. *APCO*

9 CHRONOLOGY

Early 1977
Boeing unveils latest iteration of 'Stretched 727' design, to be powered by Pratt & Whitney JT10D-4 (later the PW2037), Rolls-Royce RB211-535 or General Electric CF6-32.

Mid 1977
Boeing 7N7 new airframe with T-tail with up to 180 seats.

February 1978
Boeing 7N7 became the 757.

31 August 1978
Commitment for 19 757s from BA and 21 757s for Eastern.

23 March 1979
Start of engineering of 757.

April 1979
First run of RB211-535 turbofan on test-bed.

10 December 1979
First metal cut.

12 November 1980
Delta Air Lines ordered 757s and selected PW2037 engines.

January 1981
Assembly commenced at Renton.

December 1981
PW2037 turbofan first run on test-bed.

23 January 1982
Rolls-Royce RB211-535C turbofans run for first time on 757.

19 February 1982
Boeing 757 N757A maiden flight from Renton.

17 December 1982
US Federal Aviation Administration (FAA) certification achieved for Boeing 757.

22 December 1982
First 757 delivered to Eastern.

1 January 1983
757 operates first commercial service, for Eastern Airlines.

7 January 1983
757 enters regular commercial service with Eastern Airlines, on Atlanta-Tampa route.

January 1983
UK Civil Aviation Authority (CAA) type certification awarded.

25 January 1983
First 757 delivered to British Airways.

9 February 1983
British Airways starts 757 commercial operations, on London Heathrow-Belfast Shuttle.

September 1983
Cat IIIB autoland certificated with British Airways.

30 November 1983
RB211-535-E4 turbofan receives type certificate.

3 February 1984
RB211-535E4 engine powers 757 for first time.

14 March 1984
First flight of 757 powered by PW2037.

29 October 1984
First customer delivery of RB211-535E4-powered 757-200, to Eastern Airlines.

30 May 1985
First 757-200ER ordered by Royal Brunei.

31 December 1985
Boeing 757PF launched by order from UPS.

17 February 1986
Royal Nepal Airlines placed order for 757-200 Combi.

13 June 1986
100th 757 delivered, to Royal Brunei.

RIGHT: The No.1 Boeing 757 (N757A) was used for the more demanding aspects of the flight testing and remains the property of Boeing. The flight development programme was shared with four other 757s destined for delivery to British Airways and Eastern Airlines in the 10-month long certification programme. *Boeing*

CHRONOLOGY

ABOVE: The prototype Boeing 757, 'Ship No. 1' has remained in Boeing hands and has subsequently performed a wide range of test and trials tasks, not all necessarily linked to the 757 programme itself. *Boeing*

6 October 1987
First 757PF delivered to UPS.

November 1987
The Mexican Air Force becomes the first military 757 operator, accepting a VIP configured aircraft for service.

22 November 1988
200th 757 delivered, to China Southern.

24 August 1990
300th 757 delivered, a 757PF to Ethiopian.

18 January 1991
Eastern Airlines ceased operations.

19 October 1991
400th 757 delivered, to TAESA.

12 November 1992
500th 757 delivered, to Delta.

15 February 1994
600th 757 delivered, to Britannia Airways.

25 March 1996
700th 757 delivered, to Northwest.

8 August 1996
US Air Force selects 757 as the winner of its VC-X competition. Orders four aircraft to be known as C-32s.

September 1996
Stretched 757-300 launched by commitment from Condor.

November 1997
Assembly began of first 757-300.

11 February 1998
First flight of C-32 for USAF.

3 April 1998
800th 757 delivered, to Delta.

2 August 1998
Maiden flight of 757-300.

January 1999
757-300 awarded EAA and JAA certification.

10 March 1999
First 757-300 delivered to Condor.

2 December 1999
900th 757 delivered, to Iberia.

2 May 2000
TWA places the order for the 1,000th 757 as part of a deal for 20 aircraft.

4 May 2000
American Trans Air becomes the North American launch customer for the 757-300, with an order for 10 aircraft.

11 May 2000
jmc Airlines places first UK order for 757-300, two aircraft.

ABOVE: The Boeing 757 Freighter destined for Ethiopian Airlines started final assembly at Renton in June 1990. It was the 300th 757 and was delivered on 24 August 1990. *Boeing*

BELOW: Assembly of the first stretched 757-300 eventually destined for Condor started at Renton in March 1998 with the front fuselage section ready for joining to the centre-section, and the rear fuselage being lifted into place. *Boeing*

INDEX

Accidents, 102-3
Air 2000, 31, 54, 68, 105-8, 112, 115, 121
Air China, 61
Air Europe, 50, 52, 57, 60, 74, 104-5, 108-10, 112
Air Florida, 48
Air Holland, 53, 54, 55, 74, 107-9
Air Malta, 48
air JMC, 31
Airbus, 10, 17, 50, 52, 56, 64
Airtours, 69, 104, 109, 119
Aloha, 21
American Airlines, 21, 31, 48, 55, 65, 70, 102, 109-19, 121-2
American Trans Air, 31, 53, 71, 115, 118-23, 126
American West Airlines, 52, 53, 56, 66-7, 104-7, 109
Ansett Worldwide Aviation, 28, 69
Automation, 19
Avianca, 58, 72-3, 91, 116

Boeing 7N7, 9-10
Boeing 707, 9, 53, 62, 100
Boeing 727, 9, 10, 13, 21, 24, 34, 61
Boeing 737, 31, 32, 37, 64
Boeing 747, 9
Boeing 757-200 Combi, 26-9, 38, 93, 124; 757-200ER, 38, 52, 55 ; 757-200PF, 38; 757-200X, 31; 757-300, 29, 31, 37, 38, 62, 126
Boeing 767, 11, 17, 18-19, 21-2, 24, 41
Brakes, 24, 28, 45, 46,
Britannia Airways, 2-3, 45-7, 56, 59, 74, 103, 110, 112-6, 118, 122, 126
British Airways, 10, 12, 14-15, 16, 21, 25, 48, 49, 50, 51, 64, 75, 80, 84, 104-9, 110, 112, 116-22, 124

CAAC, 54-6
Canada 3000, 68, 76, 106-8, 110-11, 119
Cargo, 10, 26-9, 38, 48, 64-5, 76-7, 80, 98-9
Certification, 24, 25, 26, 30, 31, 40, 48, 50, 58, 124, 126
China Southern Airlines, 76-7, 107-14, 126
Composites, 16, 32-3, 46
Computers, 44, 102
Condor, 29, 30, 31, 36-7, 62, 65, 78, 110, 112-5, 121-3, 126-7
Continental Airlines, 31, 56, 59-60, 62, 78-9, 116-23

Costs, 10, 15, 17, 34, 37, 64

Delta AirLines, 21, 24, 26, 48-9, 58, 78-9, 104-15, 118-9, 121-4, 126
Deregulation, 11
Design and development, 16-31
DHL, 48, 58, 64, 80, 109

Eastern Airlines, 10, 12, 16, 21, 26, 48, 49, 50-1, 56-8, 67, 69, 104-6, 124, 126
El Al, 53, 55, 80, 107-8, 110-11, 115, 119
Electronics, 20, 24, 43-45, 46-7
Ethiopian Airlines, 28, 57, 61, 81, 110-14, 126-7
ETOPS, 26, 28, 29, 31, 58
Evolution, 8-15

Far Eastern Air Transport, 62-3, 112, 122-3
Finnair, 49, 81-2, 119-22
Flight deck, 13, 15, 24, 41, 43-4
Fuel, 9, 16-17, 19, 20-1, 25, 28, 31-2, 38, 41, 46-7, 49, 56,
Fuselage, 9, 13, 14, 15, 17, 29, 30-2

GPA, 60, 68, 73, 5

Hydraulics, 47

Iberia, 56, 64, 83, 104, 110, 115-6, 120, 122-3, 126
Icelandair, 55-6, 63, 83, 109-12, 118, 120, 122
ILFC, 50, 52, 54, 56, 60, 61, 62

jmc AIR/Flying Colours/Caledonian, 54, 57, 60-1, 63, 84-5, 108, 112, 114-5, 118-9, 121-2, 126

Ladeco, 59, 112-13
LTU, 50-1, 53, 86, 104-5, 108, 118, 122

McDonnell Douglas, 17, 48
Military use, 33, 56, 62, 64-5, 100-1, 120-21, 126
Monarch Airlines, 48, 50-2, 87, 104, 106-8

NASA, 56, 104
North American Airlines, 88-9, 117, 122

Northwest Airlines, 24, 51, 60, 90, 105-8, 118, 126
Nose, 12, 15, 43

Painting, 37
Partners/sub-contractors, 17, 32-3, 39, 65
Passenger capacity, 9-11, 13, 15, 17, 29, 38, 42, 48-9, 52, 62, 64
Payload, 10-11
Powerplant, 9, 10, 12-13, 14-15, 20-1, 24, 28-9, 31, 38-43, 48-9, 62, 124
Privatair, 90-1
Production, 14-15, 32-37

Range, 10, 25-6, 29, 32, 33, 38
Royal Air Maroc, 52, 92-3, 106
Rudder, 16-7, 29, 31, 33, 44

Sales/orders, 31, 37, 48-9, 50-8, 62-3, 65
Shanghai Airlines, 54, 58, 94-5, 108-10, 116-8
Singapore Airlines, 24, 50, 53, 71
Speed, 12, 18, 23, 29, 30, 38, 45

Taesa, 60, 67, 109, 112, 114, 126
Tail, 9, 10, 13, 14, 16, 24, 31, 33, 124
Testing, 14, 19-21, 23-4, 29-31, 34, 39-40, 42, 55, 126
Trans World Airlines, 31, 61, 63, 96-7, 119-23, 126
Transavia, 59, 65, 94-5, 114-6, 118,
Transbrasil, 21

United Airlines, 9-10, 12, 55, 59, 61-2, 97-8, 109-16, 118-22
United Parcel Service, 26-8, 42, 58, 98-9, 107-8, 110-12, 114, 116-22, 124, 126
US Airways/US Air, 57, 99, 104, 115-8
USAF, 56, 62, 64-5, 100-01, 120-21, 126

Weights, 9, 12, 15-18, 25-7, 29-30, 32-3, 38, 42, 46
Wheels, 43, 44, 47
Wings, 9, 13-4, 16, 18-19, 29, 32-3, 37-8, 43, 46

Xiamen Airlines, 58, 100-1, 103, 114, 116, 118

Zambian Airways, 28